QUICK LOOK
GENETICS

QUICK LOOK GENETICS

Patricia A. Hoffee, Ph.D.

Professor of Molecular Genetics and Biochemistry
Department of Molecular Genetics and Biochemistry
University of Pittsburgh School of Medicine
Pittsburgh, Pennsylvania

Fence Creek Publishing

Madison, Connecticut

Production by Karen Feeney, Derra
Typeset by Pagesetters, Brattleboro, VT
Printed by Port City Press, Baltimore, MD
Illustrations by Visible Productions, Fort Collins, CO
Cover design by Calvin Nelson

Distributors:

USA

Blackwell Science, Inc.
Commerce Place
350 Main Street
Malden, Massachusetts 02148
(Telephone orders: 800-215-1000 or 781-388-8250; fax orders: 781-388-8270)

Canada

Login Brothers Book Company
324 Saulteaux Crescent
Winnipeg, Manitoba, R3J 3T2
(Telephone orders: 204-224-4068)

Australia

Blackwell Science Pty, Ltd.
54 University Street
Carlton, Victoria 3053
(Telephone orders; 03-9347-0300;
fax orders: 03-9349-3016)

Outside North America and Australia

Blackwell Science, Ltd.
c/o Marston Book Services, Ltd.
P.O. Box 269
Abingdon
Oxon OX14 4YN
England
Telephone orders: 44-01235-465500;
fax orders: 44-01235-465555)

Printed in the United States of America
99 00 01 02 5 4 3 2 1

Contents

Preface

After more than 30 years of teaching genetics to medical students, I have learned that it is not humanly possible for them to retain all the information they are given in their 4 years of medical school. Medical students are, however, intelligent and bright and do in fact learn tremendous amounts of material. Often it takes just a small note or a figure to trigger their memories and bring important concepts back to them in some detail. The goal of *Quick Look Genetics* is to act as that trigger. As such, in one quick glance at a composite figure and in reading a succinct description of important concepts in medical molecular genetics, the reader should be able to recall many of the details of inherited genetic diseases, including their molecular bases and their impact on the human population.

The information should serve as a quick review of *Medical Molecular Genetics* from Fence Creek's *Integrated Medical Science (IMS) Series* and assumes that the reader has previously taken a formal course in this area. The material presented should be useful in reviewing for medical examinations as well as offering a quick review of concepts that physicians may need to know in their every day practice. With the increasing emphasis on using molecular approaches to diagnose and treat human diseases it is essential that all physicians understand the basic concepts of medical molecular genetics.

Quick Look: Genetics reviews four main areas of medical molecular genetics, including: Molecular Aspects of Human Genetics, Mendelian Inheritance, Mapping and Cloning of Human Genes, and Clinical Aspects of Human Genetics. Each part contains practice questions and problems that are designed to enable the readers to test their understanding of the material.

Patricia A. Hoffee

Dedication

To little Danielle, who every day brings joy to my life.

Acknowledgments

I would like to acknowledge all of my colleagues at the University of Pittsburgh School of Medicine for their dedication to teaching and their constant efforts to make medical molecular genetics an exciting area of learning for students. In particular, I would like to thank the following individuals who generously offered their expertise in the preparation of this book:

Dr. Eric Hoffman, Dr. W. Allen Hogge, Dr. Saleem Khan, Dr. Mary Marazita, Dr. Stephen Phillips, Dr. Paul Robbins, and Dr. Urvashi Surti.

Abbreviations

A	Adenine
ADA	Adenosine deaminase
AP	Apurinic
ARS	Autonomous replicating sequence
AS	Angelman syndrome
ASO	Allele-specific oligonucleotide
ASOH	Allele-specific oligonucleotide hybridization
ATP	Adenosine triphosphate
ATPase	Adenosine triphosphatase
AUG	Initiator codon
BMD	Becker muscular dystrophy
bp	Base pairs
BRCA	Breast cancer gene
C	Cytosine
CA repeats	Repeats of cytosine and adenine
CAAT	CAAT box
Cap	7-methylguanosine
Cdc2	Mitotic cyclin-dependent kinase
Cdk	Cyclin-dependent kinase
cDNA	Complementary DNA
CF	Cystic fibrosis
CFTR	Cystic fibrosis transmembrane conductance regulator
CKI	Cyclin-dependent kinase inhibitor
cM	CentiMorgan
COOH	Carboxyl group
CS	Cockayne syndrome
CVS	Chorionic villus sampling
Cyt C	Cytochrome C
D	Daltons
dCTP	Deoxycytosine triphosphate
DMD	Duchenne muscular dystrophy
DNA	Deoxyribonucleic acid
DRPA	Dentatorubral-pallidoluysian atrophy
E1, E2, E3	Enzymes of the ubiquitin-proteasome pathway
E1B	E1B protein of adenovirus
E2F	Transcription factor
EF1	Elongation factor 1
EGF	Epidermal growth factor
ER	Endoplasmic reticulum
EST	Expressed sequence tag
FAP	Familial adenomatous polyposis
FISH	Fluorescent in situ hybridization
FMR1	Fragile X mental retardation
G	Guanine
G-banding	Giemsa chromosome banding
G1	G1 phase of the cell cycle
G2	G2 phase of the cell cycle
GTF	General transcription factor
GTP	Guanosine triphosphate
GTPase	Guanosine triphosphatase
HAT	Hypoxanthine-aminopterin-thymidine selection medium
Hb	Hemoglobin
HbA	Adult hemoglobin
HbF	Fetal hemoglobin
HGPRT	Hypoxanthine-guanine phosphoribosyl transferase
HNPCC	Hereditary nonpolyposis colorectal cancer
hnRNA	Heterogenous nuclear ribonucleic acid
HRE	Hormone response element
IF2	Initiation factor 2
IgM	Immunoglobulin M
Inr	Initiator sequence
IRP	Iron response protein
ISS	Inhibitor of sister chromatid separation
kD	kilodalton
LCR	Locus control region
LDL	Low-density lipoprotein
LHON	Leber's hereditary optic neuropathy
LINE	Long interdispersed nuclear element
LOD	Logarithm of the odds
LOH	Loss of heterozygosity
M	Mitosis
MELAS	Mitochondrial myopathy, encephalopathy, lactic acidosis, and stroke-like episodes
MERRF	Myoclonic epilepsy with ragged red fibers

MMR	Methyl-directed mismatch repair
mRNA	Messenger RNA
Mut	Mutator protein
mut	Mutator gene
MYC	*MYC* gene (oncogene)
NER	Nucleotide excision repair
NF	Neurofibromatosis
NH_2	Amino group
OI	Osteogenesis imperfecta
ori	Origin of DNA replication
p arm	Short arm of a human chromosome
p53	Protein 53
PCR	Polymerase chain reaction
PDGF	Platelet-derived growth factor
PKU	Phenylketonuria
pRb	Retinoblastoma protein
*Pst*1	*Pst*1 endonuclease
PUBS	Percutaneous umbilical blood sampling
PWS	Prader-Willi syndrome
q arm	Long arm of a human chromosome
R point	Restriction point in the cell cycle
ras	*ras* oncogene
RB	Retinoblastoma
RBC	Red blood cells
RE	Response element
RF	Release factor
RFLP	Restriction fragment length polymorphisms
RNA	Ribonucleic acid
RNAse	Ribonuclease
rRNA	Ribosomal ribonucleic acid
S	S phase (DNA replication phase)
snRNA	Small nuclear RNA
SP1	Transcription factor
SSCP	Single-strand conformational polymorphism
STRs	Short tandem repeats
STS	Sequence tagged site
SWI/SNF	Complex associated with RNA polymerase II
T	Thymine
T-T	Thymine-thymine dimer
TAF	TBP-associated factors
TATA box	*Cis*-acting sequence in RNA polymerase II promoters
TBP	TATA-binding protein
TF	Transcription factor
TFII	Transcription factor for control of RNA polymerase II
tRNA	Transfer RNA
TTD	Trichothiodystrophy
U	Uracil
U_1–U_6	snRNA molecules involved in splicing
UGA, UAA, UAG	Termination codons
UTR	Untranslated region
UV	Ultraviolet irradiation
VNTR	Variable number tandem repeats
XP	Xeroderma pigmentosum
YAC	Yeast artificial chromosome

Figure Credits

CHAPTER 3

Part B. Reprinted with permission from Becker W, Reece JB, Poenie M. *The World of the Cell*, 3rd ed, Menlo Park, CA: Benjamin-Cummings, 1996, p 434.

Part C. Reprinted with permission from Becker W, Reece JB, Poenie M. *The World of the Cell*, 3rd ed, Menlo Park, CA: Benjamin-Cummings, 1996, p 435.

CHAPTER 4

Part C. Taken from the 1971 Paris Conference on Standardization in Human Cytogenetics sponsored by the March of Dimes.

CHAPTER 7

Part A. Reprinted with permission from Becker W, Reece JB, Poenie M. *The World of the Cell*, 3rd ed, Menlo Park, CA: Benjamin-Cummings, 1996, pp 474–475.

Fig C. Reprinted with permission from Barinaga M. *A new twist to the cell cycle.* Science 269:693, 1995

CHAPTER 8

Parts A and B. Reprinted with permission from Becker W, Reece JB, Poenie M: The World of the Cell, 3rd ed, Menlo Park, CA: Benjamin-Cummings, 1996, pp 502–503.

CHAPTER 9

Parts B and C. Compliments of Dr. Stephen Phillips, University of Pittsburgh School of Medicine.

CHAPTER 20

Part C. Reprinted with permission from Weatherall DJ, Clegg JB, Higgs DR, et al: The hemoglobinopathies. In *The Metabolic and Molecular Bases of Inherited Diseases*, 7th ed. Edited by Scriver C, Beaudet A, Sly W, et al. New York, NY: McGraw-Hill 1995 p 3426.

CHAPTER 22

Part B. Adapted with permission from Lasko D, Cavenee W, Nordenskjold Ml. *Loss of constitutional heterozygosity in human cancer.* Ann Rev Genet 25:286, 1991.

CHAPTER 23

Part C. Compliments of Dr. Paul Robbins, University of Pittsburgh School of Medicine.

CHAPTER 24

Parts A, B, and C. Compliments of Dr. Stephen Phillips, University of Pittsburgh School of Medicine.

CHAPTER 25

Part A. Adapted with permission from Bennett RL, Steinhaus KA, Uhrich SB et al. *Recommendations for standardized human pedigree nomenclature.* Am J Hum Genet 56;745–752, 1995.

CHAPTER 27

Part B. Reprinted with permission from Tsui LC, Buchwald M. Biochemical and molecular genetics of cystic fibrosis. *In Advances in Human Genetics*, vol 20. Edited by Harris H, Hirschhorn K. New York, NY: Plenum Press 1991, p 208

Part C. Reprinted with permission from Grabowski GA. *Gaucher's Disease in Advances in Human Genetics*, vol 21. Edited by Harris H, Hirschhorn K. New York, NY: Plenum Press, 1993, p 404.

Part D. Adapted with permission from Scriver C, Kaufmann S, Eisensmith R, et al. The hyperphenylalaninemias. In *The Metabolic and Molecular Bases of Inherited Diseases*, 7th ed. Edited by Scriver C, Beaudet A, Sly W, et al. New York, NY: McGraw-Hill 1995 p 1028.

CHAPTER 31

Part A. Adapted with permission from Nicholls AC. Collagen disorders. In *The Inherited Metabolic Diseases.* Edited by Holton J, London, UK: Churchill Livingston, 1994, p 381.

CHAPTER 32

Part A. Reprinted with permission from Hobbs HH, Russell D, Brown M, et al. *The LDL receptor locus and familial hypercholesterolemia.* Ann Rev Genet 24: 136, 1990.

Parts B and C. Reprinted with permission from Hobbs HH, Russell D, Brown M, et al. *The LDL receptor locus and familial hypercholesterolemia.* Ann Rev Genet 24:141, 1990.

CHAPTER 33

Part B. Adapted with permission from Malkin D, Li F, Strong J, Fraumeni J, et al. *Germ line p53 mutations in a familial syndrome of breast cancer, sarcomas and other neoplasms.* Science 250:1234, 1990.

CHAPTER 35

Fig A. Reprinted with permission from Kahn P. *Coming to grips with genes and risk.* Science 274:496, 1996.

CHAPTER 38

Part B. Adapted with permission from Johns D. *The other human genome: mitochondrial DNA and disease.* Nature Med. 2:1066, 1996.

CHAPTER 43

Part B. Compliments of Dr. Stephen Phillips, University of Pittsburgh School of Medicine.

CHAPTER 47

Part A. Reprinted with permission from Shows TB, Sakaguchi AY. *Gene transfer and gene mapping in mammalian cells in culture*. In Vitro 16:65, 1980.

Part B. Reprinted with permission from Gray JW, Langlois RG. *Chromosome classification and purification using flow cytometry and sorting*. Ann Rev Biophys Biophys Chem 15:224, 1986.

CHAPTER 48

Part B. Reprinted with permission from Murray JC, Buetow KH, Weber JL et al. *Human genetic map-genome map*. Science 265: 2055, 1994.

CHAPTER 49

Part C. Reprinted with permission from Schwartz LS, Tarleton J, Popovich B, et al. *Fluorescent multiplex linkage analysis and carrier detection for Duchenne/Becker muscular dystrophy*. Am J. Hum Genet 51:727, 1992.

Part C. Reprinted with permission from Miller RG, Hoffman EP. *Molecular diagnosis and modern management of Duchenne muscular dystrophy*. Neurologic Clinics 12:715, 1994

CHAPTER 63

Part B. Reprinted with permission from Curiel D. *Adenovirus facilitation of molecular conjugate-mediated gene transfer*. Prog Med Virol. 40: 9–12. 1993

CHAPTER 64

Part B. Compliments of Dr. Paul Robins, University of Pittsburgh School of Medicine and the University of Pittsburgh Medical Center.

PART I
Molecular Aspects of Human Genetics

1 DNA Structure

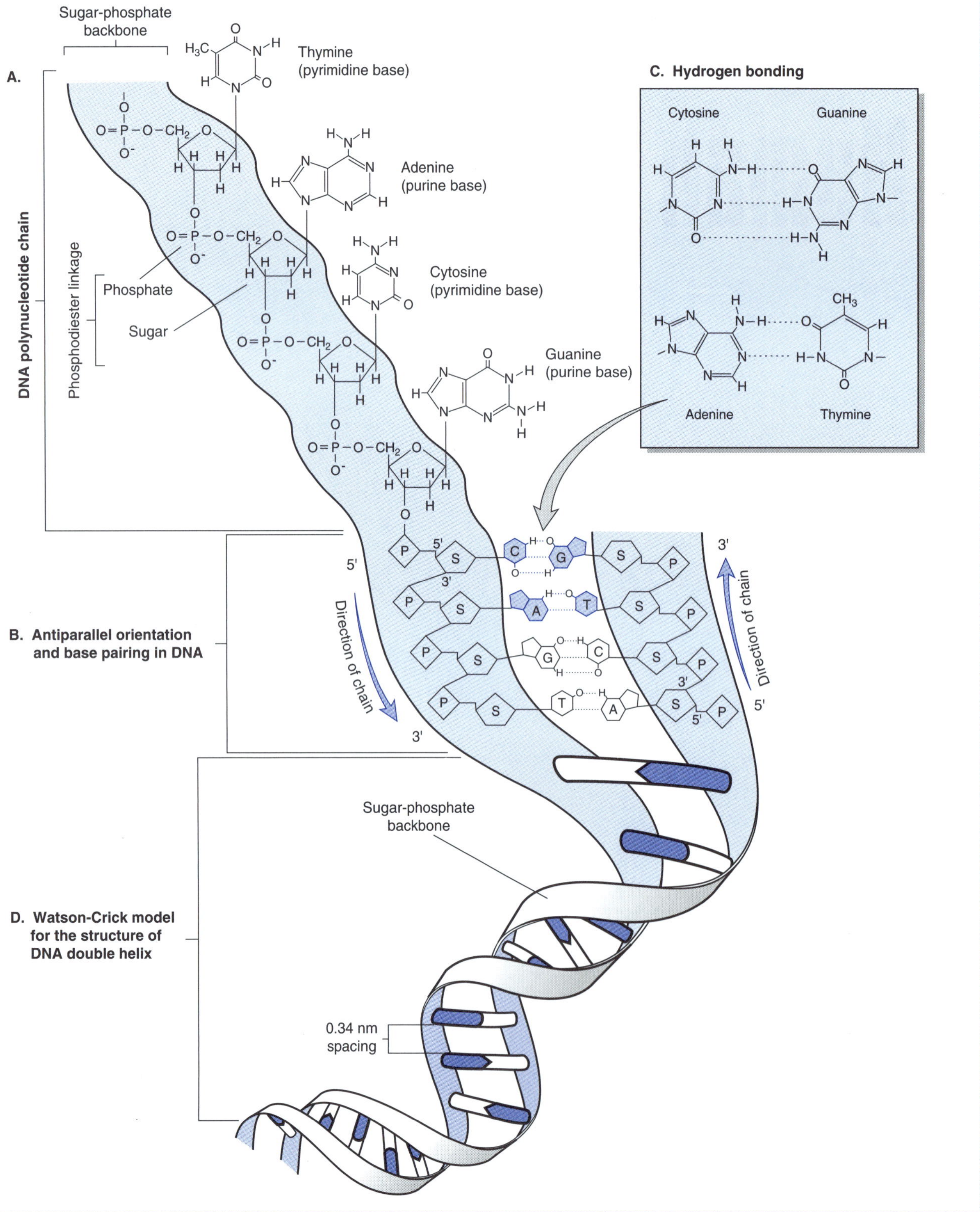

OVERVIEW

Human cells, along with most organisms, store their genetic information in molecules of deoxyribonucleic acid, or DNA. The information stored in a DNA molecule must be stable, must be resistant to change, and must be faithfully replicated. All of these properties are important because they prevent the occurrence of mistakes or defects arising in the DNA and being transmitted to daughter cells. Additionally, the sequence of bases in a DNA molecule determines whether a particular RNA or protein is expressed during the growth and life of a cell. These features of DNA—stability, resistance to change, faithful replication, and the ability to transfer the information contained in its sequence to the cell—are built into the structure of the molecule.

Polynucleotide Chains

All DNA molecules, whether they are found in prokaryotic or eukaryotic cells, are made up of four types of nucleotides linked together by phosphodiester bonds to form long chains or polynucleotides (Figure **A**). Each nucleotide in DNA consists of a nitrogenous base, the sugar deoxyribose, and a phosphate residue. The nitrogenous bases found in DNA are of two types: purines (i.e., adenine [A] and guanine [G]) and pyrimidines (i.e., cytosine [C] and thymine [T]). These nucleotides are linked by covalent phosphodiester bonds that join the 5′ carbon of one deoxyribose to the 3′ carbon of the next deoxyribose to form polynucleotide chains.

Antiparallel Orientation and Base Pairing in DNA

The DNA chain can exist as a single-stranded molecule but is found most often in cells as a double-stranded structure in which two polynucleotide chains are held together by hydrogen bonding between purine and pyrimidine bases. This structure is helical in nature, with two strands of DNA coiled around a common axis and running in antiparallel (opposite) orientation. Opposite orientation, or polarity of the molecule, is an important concept in the biochemistry of DNA. Figure **B** illustrates the polarity of two strands of DNA in the helix with the bases hydrogen-bonded in a complementary manner. One DNA strand has a 5′ to 3′ orientation, while the complementary strand runs in a 3′ to 5′ direction. This directionality will also be seen in RNA molecules and is important to remember for an understanding of many of the functions RNA and DNA play in the cell.

Hydrogen Bonding: Adenine and Thymine, Guanine and Cytosine

When purines and pyrimidines are face to face in the double-stranded helix, their pairing properties result in an adenine (A) always pairing with a thymine (T) and a guanine (G) always pairing with a cytosine (C). The DNA helix is made up of successive A-T and G-C base pairs along the length of the polynucleotide chains. The two strands of the double helix are complementary: that is, A always hydrogen bonds to T, and G always hydrogen bonds to C (Figure **C**). Therefore, given the sequence of bases on one strand of the DNA double helix, the sequence on the complementary strand is fixed and easily determined. For example, the sequence ACCTGGACATTG on one strand of DNA will have the sequence TGGACCTGTAAC on the complementary strand of DNA. The variation of the sequence of nucleotides along the DNA strand determines the function of each section of the DNA molecule and its ability to transmit information to RNA and protein.

Watson-Crick Model for the Structure of the DNA Double Helix

The most common form of DNA in nature is the B-DNA structure. This DNA double helix structure was determined by Watson and Crick in 1953 (Figure **D**). In the B-DNA structure, the deoxyribose-phosphate backbone is located on the outside of the molecule, and the bases are stacked perpendicular to the axis of the helix inside the molecule. The two strands of DNA, which wrap around each other in a right-handed turn, are held together by hydrogen-bonded bases, which are face to face inside the molecule. A simple way to visualize the helix is to imagine a rope ladder with the rungs of the ladder formed by the base pairing A-T or G-C. If the ladder is given a right-handed twist, the result is a helix in the form of B-DNA. In B-DNA, each turn of the helix represents 10 base pairs (bp) and is equivalent to 3.4 nm in length. This number is fixed, enabling us to calculate the length of any B-DNA molecule from the number of base pairs it contains.

The total DNA content of a cell makes up the entire genetic information available to the organism and is referred to as its genome. DNA genomes come in various sizes, with the human genome containing approximately 3.3×10^9 total base pairs per haploid nucleus—more than 1000 times that found in a bacterial cell. DNA molecules can exist in several different forms in cells. For example, bacterial cells and mammalian mitochondria have DNA molecules that are double-stranded, covalently closed, circular molecules. Some mammalian viruses, such as adenovirus, have linear double-stranded DNA molecules. Human cells, with the exception of the sperm and the egg, contain 46 molecules of linear double-stranded DNA in their nuclei, each of which is combined with protein molecules to make up bodies called chromosomes. A normal human cell contains 23 pairs of chromosomes: 22 autosome pairs and 1 pair of sex chromosomes.

Clinical Significance

The sequence of nucleotides in a DNA molecule determines the sequence of amino acids in each of the proteins present in the human body. A change in the nucleotide sequence in the DNA—even the change of a single base pair in 3.3×10^9 total base pairs—can result in an alteration of the structure of a protein and lead to a loss or alteration of its function. All inherited human diseases result from changes in the nucleotide sequence of DNA.

For more information see Hoffee PA: *IMS: Medical Molecular Genetics*. Fence Creek, pp. 17–22.

2 RNA Structure

A RNA polynucleotide chain formed by 3'–5' phosphodiester bonds

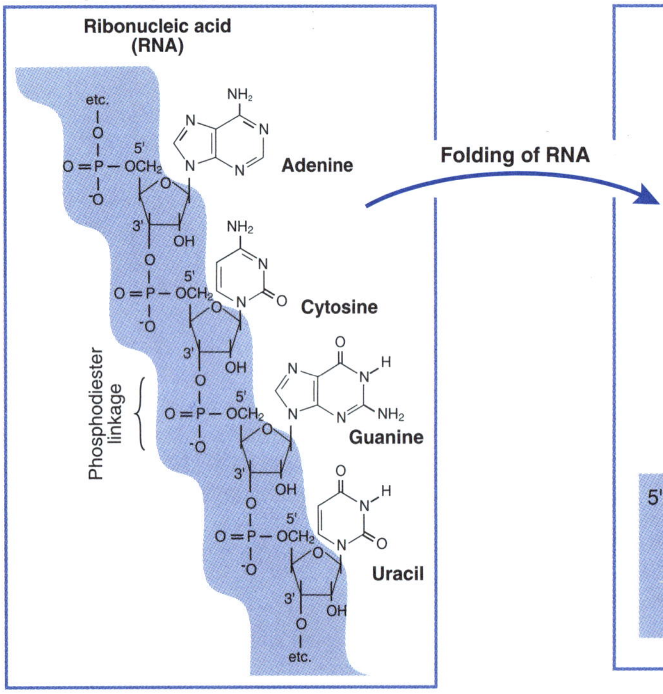

B Intramolecular hydrogen bonding in tRNA molecules

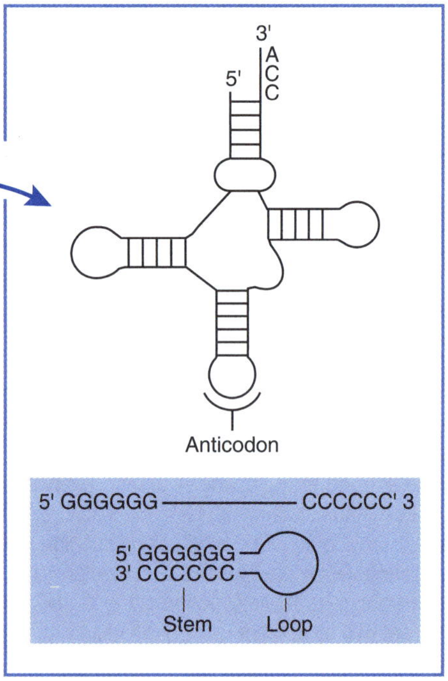

C Classes of RNA molecules found in human cells

RNA Class	*Size*	*Function*
hnRNA	Heterogeneous	Precursors to other RNAs
snRNA	58–220 nucleotides	RNA processing
mRNA	Heterogeneous	Carry information to the protein synthesis machinery
tRNA	74–95 nucleotides	Transfer amino acids
rRNA	28S, 18S, 5.8S, 5S	Make up the ribosome structure

D Information transfer in cells

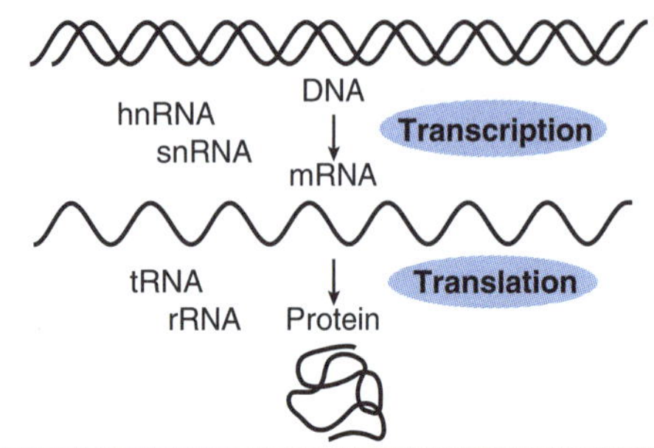

OVERVIEW

RNA molecules are transcribed directly from DNA molecules and are present in the nucleus, the cytoplasm, and the mitochondria. Although some viruses, such as retroviruses or poliovirus, contain RNA as their genetic material in place of DNA, the major function of RNA molecules in human cells is their involvement in the transfer of information from DNA to protein.

RNA Polynucleotide Chain Formed by 3′–5′ Phosphodiester Bonds

Ribonucleic acid (RNA) molecules, like DNA molecules, are composed of nucleotides linked together by phosphodiester bonds, but they generally occur as single-stranded polynucleotides and are usually shorter than most DNA molecules (Figure **A**). As with DNA, RNA is made up of the bases A, G, and C, but it contains uracil (U) in place of thymine (T). The other major difference between RNA and DNA is the presence of the sugar ribose in RNA in place of the deoxyribose found in DNA. Since U has the ability to pair with A in much the same way as T pairs with A, the four bases found in RNA—A, U, G, and C—can form complementary pairs with other bases in RNA as well as with the bases found in DNA.

Intramolecular Hydrogen Bonding in tRNA Molecules

Because of the single-stranded nature of RNA and the pairing properties of the four bases, RNA often contains intramolecular hydrogen bonding and secondary structures (Figure **B**). An important point to notice is the directionality of the RNA molecule, going in a 5′ to 3′ direction. If the molecule twists around itself, opposite polarity can occur in some regions of the molecule, resulting in the formation of base pairs between complementary purine and pyrimidine bases. This intrastrand base pairing allows a single-stranded molecule of RNA to form double-stranded portions. Such base pairing regions give structures known as stem-loop structures, with the base pairing sections forming the stem and the noncomplementary bases forming the loop. Intramolecular hydrogen bonding gives RNA a secondary structure as well as a tertiary structure, which makes the molecule very compact, more stable, and able to fit in a smaller space. In addition, the presence of a secondary structure allows the RNA molecule to carry out biologic functions more efficiently than a linear molecule. This property of RNA is especially important in the process of protein synthesis.

Classes of RNA Molecules Found in Human Cells

Five main classes of RNA are found in human cells. They include heterogeneous nuclear RNA (hnRNA), small nuclear RNA (snRNA), messenger RNA (mRNA), transfer RNA (tRNA), and ribosomal RNA (rRNA) (Figure **C**).

hnRNA and snRNA.

These molecules are found in the nucleus of the human cell. hnRNA is the immediate product of transcription, is complementary to one strand of the DNA, and is the precursor to mRNA before it is processed. snRNA is found associated with specific proteins and is involved in the processing of the hnRNA to mRNA before the exit of mRNA from the nucleus to the cytoplasm.

mRNA.

mRNA makes up approximately 2%–5% of the total RNA in the cell, is short lived, and shows a large variation in base sequence from one mRNA molecule to another. These are the messengers that carry the information from the DNA to the protein-synthesizing machinery.

tRNA.

tRNA molecules are small polynucleotides ranging from 74 to 95 nucleotides in length. Their major role is to carry specific amino acids to the ribosomes during protein synthesis. For each of the 20 amino acids found in proteins, there is a unique tRNA molecule, which specifically recognizes that amino acid. In some cases there is more than one tRNA species for a single amino acid.

rRNA.

rRNA is the most abundant RNA species in cells, making up nearly 80% of the total, and is found associated with proteins in structures called ribosomes. rRNA varies in size; the specific rRNAs of eukaryotic cells are designated by sedimentation coefficients (S values). Human ribosomes contain rRNA species of the following sizes: 28S, 18S, 5.8S, and 5S.

Information Transfer in Cells

Each of the different RNA molecules plays a specific role in the process of transcription (the transfer of information from DNA to RNA) or the process of translation (the transfer of information from mRNA to protein) (Figure **D**). The details of these reactions are given in Chapters 9–12.

Clinical Significance

RNAs function in a variety of steps during the transfer of information from DNA to protein, each type of RNA taking part in a specific reaction. Each of these reactions must be completed in a precise manner to ensure the accurate transfer of information. As discussed later, alterations in the base sequence of RNA or mistakes in the processing or translating of RNA often lead to altered protein structures and human disease.

For more information see Hoffee PA: *IMS: Medical Molecular Genetics*. Fence Creek, pp. 19–23.

3 Chromatin and Chromosome Structure

A Mechanisms for Compacting DNA

1. **Bending DNA molecules** at the sequence AAAAAXXXXXAAAAAXXXXXAAAAA, where A represents adenine residues and X represents any other base
2. **Supercoiling of DNA,** a process in which the double-stranded helix falls back on itself to give a less relaxed and more compact DNA molecule
3. **Forming of a complex,** consisting of DNA and basic proteins called histones

B Nucleosomes

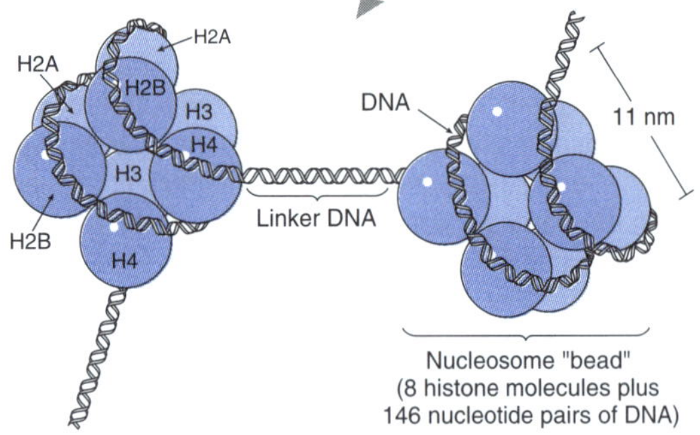

C Packaging of DNA in the nucleus

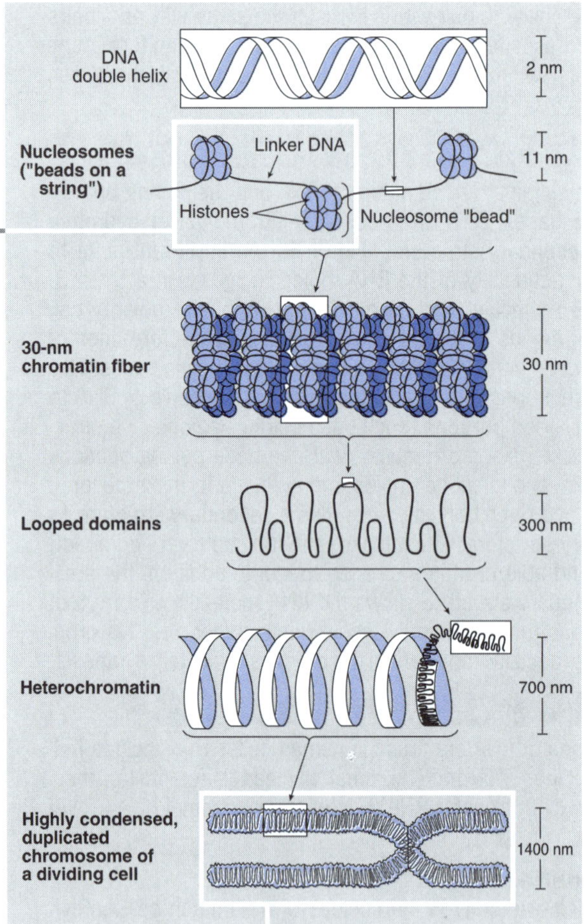

D Important elements found in human chromosomes

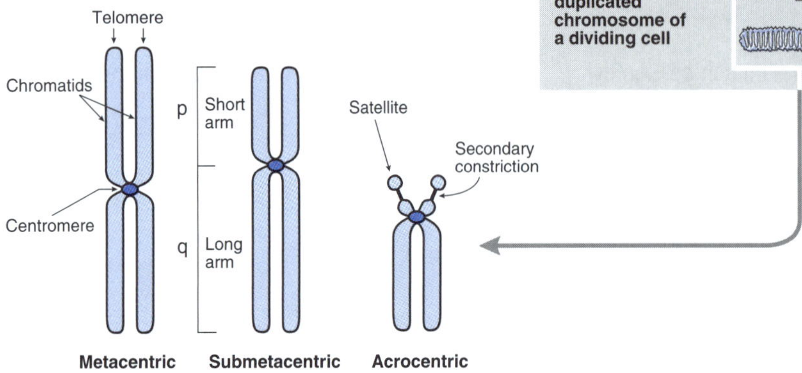

OVERVIEW

The 46 chromosomes of a human cell are found inside a small organelle called the nucleus. If we consider the number of base pairs of DNA in a human nucleus (6.6×10^9) and the fact that these base pairs are arranged into 46 individual DNA molecules, it becomes obvious that the DNA would not be able to fit into the nucleus if it remained as linear DNA. Thus, we find that the DNA molecules inside the nucleus are compacted into smaller forms by the three mechanisms shown in Figure **A**.

Nucleosomes

At all stages of the cell cycle, DNA is associated with specific basic proteins called histones and forms structures known as nucleosomes. Five major classes of histones are found associated with DNA in the nucleus: H1, H2A, H2B, H3, and H4. Histones contain high concentrations of lysine and arginine, which give them a positive charge and promote the formation of strong ionic bonds with negatively charged DNA. Under an electron microscope, chromatin is seen as repeating units of nucleosomes arranged with regular spacing to form long fibers with bead-like structures (Figure **B**). Each nucleosome is made up of a core of eight histone proteins (two copies each of H2A, H2B, H3, and H4) with 146 bp of DNA wrapped around them to form a DNA-histone complex. Adjacent nucleosomes are separated from each other by a stretch of 50 to 60 nucleotides of DNA, referred to as spacer DNA or linker DNA. The linker DNA is generally considered to be part of the nucleosome, giving a total of approximately 200 bp of DNA per nucleosome.

Packaging of DNA in the Nucleus

Histone H1 is not part of the nucleosome structure but is important in assembling multiple nucleosomes into a complex fiber structure, the 30-nm chromatin fiber, sometimes called the solenoid (Figure **C**). Additional packaging of the DNA occurs with the formation of domains, or loops of chromatin held together by nonhistone proteins. These loops, containing 50,000–100,000 bp, are attached to a scaffold structure and represent regions of less compacted DNA, known as euchromatin, which for the most part are transcriptionally active. Other areas of the DNA-protein structures are more highly compacted and make up heterochromatin, which are areas of DNA sequence believed to be transcriptionally inactive. Further compaction of the loop-scaffold DNA-protein structure produces the chromatids, which make up the chromosome structures seen only in the mitotic phase of the cell cycle, when the chromosomes become highly condensed.

Important Elements Found in Human Chromosomes

The chromosomes, as seen in mitosis, have three well-defined sequences or structures that are essential for their replication and distribution to daughter cells (Figure **D**). These elements are as follows:

DNA Replication Origins.
These are multiple sites that exist within the chromosomes where DNA replication initiates.

Telomeres.
Telomeres are important structures associated with the ends of all human chromosomes and consist of multiple tandem repeats of a specific sequence, TTAGGG, located on both ends of each chromatid. Telomeres serve several very important roles in human cells:

- They are essential for maintaining the stability of the chromosome and preventing end-to-end fusions from occurring.
- Telomeric sequences protect the ends of the chromosomes from being degraded by nucleases. Proteins appear to interact with the telomeric DNA sequences to give a nucleoprotein protective cap on the ends of each chromatid, making the DNA much more resistant to enzymatic degradation.
- Telomeres are essential for assuring the proper replication of the ends of the chromosomes.
- Telomeres play a role in normal cell growth control. The presence of a telomere on the end of a chromosome serves as a signal that a chromosome is normal. When a broken chromosome or a fragment of a chromosome arises in a cell, this abnormal structure is missing a telomeric sequence on at least one end. DNA damage controls, present in all normal human cells, recognize the abnormal chromosomal end and signal a temporary stopping of the cell cycle to allow the DNA repair systems to repair the damage. In the event the damage cannot be repaired, the cell may be directed to a programmed death.

Centromeres.
Centromeres are highly repetitive DNA sequences that comprise the site where the two sister chromatids are attached. The constriction produced is used to define the arms of the chromosome as the short (p, or petite) arm and the long (q) arm. The function of the centromere is to assure equal distribution of each chromosome to the daughter cells at cell division. Centromeres are positioned at different sites on different chromosomes and are used to differentiate the chromosome structures seen in mitosis as metacentric, submetacentric, and acrocentric.

Clinical Significance

Each time the DNA undergoes replication in a normal human somatic cell, 50–100 bp of the terminal telomeric sequences of a chromosome are lost. This shortening of the chromosome ends can be prevented by the enzyme telomerase, which normally is present in embryonic stem cells and germ cells. However, since most normal somatic cells do not appear to have an active telomerase, as a somatic cell continues to grow and divide, there is a constant shortening of the ends of the chromosomes. The continual shortening of chromosomes may function as a mitotic clock, signaling the occurrence of programmed death in aging cells before a chromosomal abnormality arises and is passed to daughter cells. The absence or shortening of telomeres also may serve as a signal to damage control systems that a cell is no longer healthy and should be eliminated before it can develop a mutation. Some cancer cells can escape the above control system and become immortalized. The telomeric sequences on the chromosomes of these cancer cells are short but remain stable in length, resulting in uncontrolled growth of a cell that is now no longer subject to damage control. The stability of the telomeres in cancer cells is apparently due to the reappearance of telomerase activity.

For more information see Hoffee PA: *IMS: Medical Molecular Genetics*. Fence Creek, pp. 76–82.

4 Cytogenetics

A Symbols used in cytogenic nomenclature

Symbol	*Description*
p	Short arm of chromosome
q	Long arm of chromosome
ace	Acentric
cen	Centromere
del	Deletion
der	Derivative chromosome
dic	Dicentric chromosome
dmin	Double minute chromosome
dup	Duplication
fra	Fragile site
hsr	Homogeneously staining region
i	Isochromosome
ins	Insertion
inv	Inversion
(–)	Loss of genetic material
(+)	Gain of genetic material
r	Ring chromosome
rcp	Reciprocal
rob	Robertsonian translocation
(;)	Semicolon; separates chromosome and chromosome regions when structural rearrangements involve more than one chromosome
t	Translocation
ter	Terminal
pter	End of the short arm
qter	End of the long arm

B G-banded human chromosomes

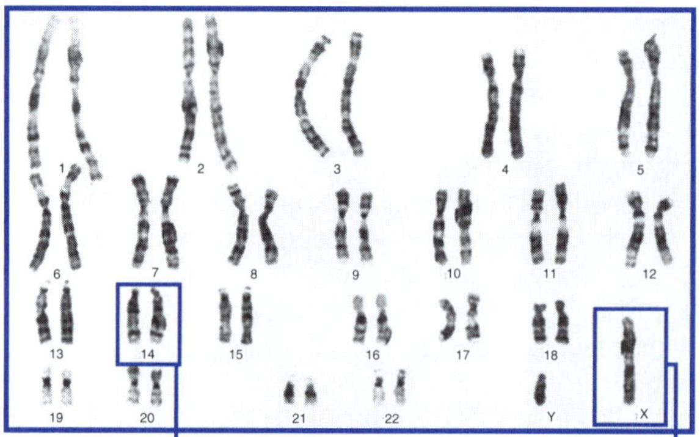

C Standard human karyotype

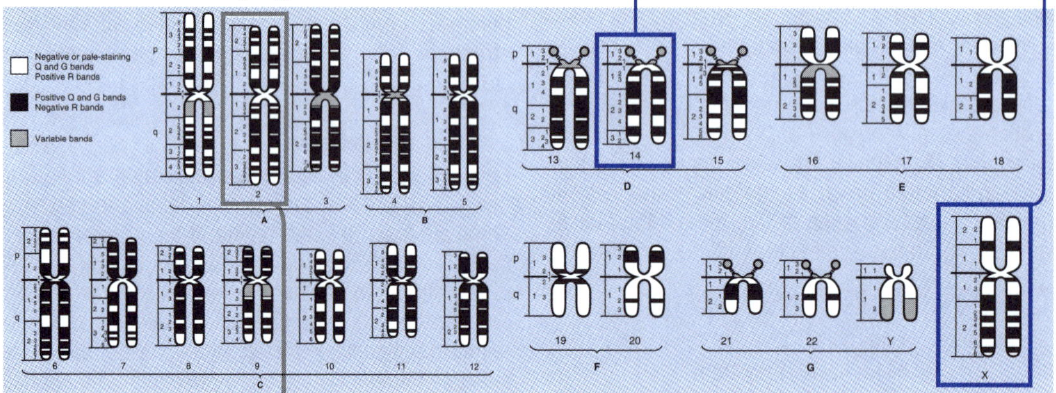

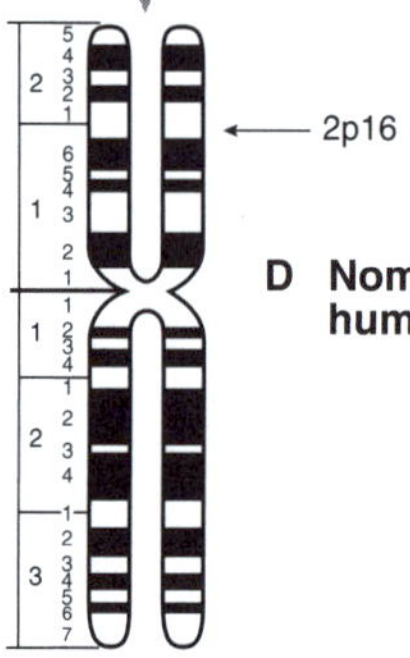

D Nomenclature for banded human chromosome

OVERVIEW

Although the 46 chromosomes found in the human nucleus are unique, some of them resemble each other when stained by certain conventional methods. Before 1971, human chromosomes could be classified only by size. In a standardized karyotype, the 22 autosome pairs were ordered by placing them in seven groups (A through G) in descending order of size. In 1971, a conference was held in Paris to establish a standardized system of nomenclature that would identify each chromosome as well as the regions of each chromosome. This system was made possible by Caspersson's discovery that when certain fluorescent dyes (e.g., quinacrine) are applied to a chromosome preparation, alternating light and dark bands result along each chromosome. Each chromosome pair gives a specific banding pattern, which makes it possible to identify accurately every chromosome in a karyotype.

Symbols Used in Cytogenetic Nomenclature

Formal standardized nomenclature for human chromosomes is maintained by a standing committee that publishes an International System of Human Chromosome Nomenclature (ISCN) to provide investigators with the accepted rules for noting karyotype designations. Some of these symbols are listed in Figure **A**.

G Banding of Human Chromosomes

Following the discovery of Q or quinacrine banding, another procedure, Giemsa banding (G banding), was developed. This procedure combines Giemsa dye staining with trypsin treatment and produces banding on human chromosomes that is identical to the bands produced by quinacrine. Regions of the chromosomes that give fluorescence with Q banding produce dark bands or G bands when treated with trypsin and Giemsa. The dark bands or G bands produced with Giemsa and trypsin represent late-replicating heterochromatin bands that are rich in adenine–thymine pairs. Approximately 20% of the human genes that have been mapped are located within the G bands. The light bands represent areas of the chromosome that replicate early in the S phase and contain most of the sequences that code for known human proteins. A typical male G-banded karyotype is shown in Figure **B**. Karyotype analysis is performed on growing cells that come from blood samples, skin samples, fetal amniocytes, chorionic villi, and tumor cells. After cells in mitosis are blocked and the appropriate staining procedure is carried out, mitotic cells are observed under a light microscope to identify metaphase spreads that have well-separated and well-defined chromosomes. Even with sophisticated automation of karyotype analysis, the time involved to complete the analysis of a karyotype is 1 to 2 weeks. Other banding procedures can be used to reveal specific features of human chromosomes. For example, constitutive heterochromatin is detected by C banding, which stains centromeric DNA as well as the satellite regions on each chromosome. A second, and very important, technique for evaluating karyotypes is known as in situ hybridization. With the increasing availability of DNA-labeled probes and the development of new labeling procedures, it is now possible to localize specific DNA sequences directly on a chromosome spread. A nonradioactive labeling procedure using a biotin-labeled probe, which can be detected in combination with a fluorescent dye, is the current method of choice. This technique, known as fluorescent in situ hybridization (FISH), can be used to localize accurately a specific gene sequence within a chromosome. Although initially used only with metaphase chromosome spreads, FISH is now used with interphase cells. The advantage in using interphase cells is the rapidity with which the technique can provide answers to the clinician.

Standard Human Karyotype

G banding is now a standard procedure for identifying metaphase chromosomes in a human karyotype. An idiogram of the standardized G bands found for the human 22 autosomes and the sex chromosomes is shown in Figure **C**. Using the designation of p for the short arm and q for the long arm, each chromosome is further divided by regions. The bands and regions on each arm of a chromosome are numbered from the centromere to the telomere. A region is any area lying between two specified landmarks. The region adjacent to the centromere is numbered 1 in each arm; the following regions are 2, 3, and so on, going toward the telomere. Each band within a region is also numbered, once again starting proximal to the centromere and going toward the telomere. In a standard chromosome preparation, about 400 total bands are seen per haploid chromosome number. Higher-resolution banding is also now available to give 1000 or more bands. In this case, the bands are further subdivided. Bands on each chromosome are named as illustrated in Figure **D**. On the p arm of chromosome 2, there are two regions (1 and 2), and within each region are the bands (e.g., in region 1, bands are noted as 1, 2, 3, 4, 5, and 6). The arrow indicates a position on region 1 and band 6, and is noted as 2p16, meaning that the location is on chromosome 2, the p arm, region 1, band 6.

Clinical Significance

The development of chromosome banding procedures coupled with FISH allows the precise localization of human genes to specific sites within each chromosome. These are essential techniques in the mapping and isolation of human genes that, when altered, lead to inherited human diseases. Interphase FISH is useful in the prenatal diagnosis of trisomies or sex chromosome abnormalities in which the extra chromosome presents as an extra fluorescent spot in the nucleus of the interphase cell. With the standardized nomenclature, a karyotype that deviates from the normal can be clearly described anywhere in the world. The sensitivity of these methods has resulted in the identification of many of the numerical and structural chromosome abnormalities that arise in the human population (see Chapters 52–59) and has established their relationship to specific inherited human diseases.

For more information see Hoffee PA: *IMS: Medical Molecular Genetics.* Fence Creek, pp. 82–86.

5 The Cell Cycle

A Proteins Involved in regulation of the cell cycle

Cell-cycle phase	Cdk	Cyclin	CKI: KIP[a]	CKI: INK[b]
G1	Cdk4	Cyclin D	p21, p27	p15, p16
G1/S	Cdk2	Cyclin E	p21, p27	
S	Cdk2	Cyclin A	p21	
G2/M	Cdc2	Cyclin B	p21	
M	Cdc2	Cyclin B, cyclin A		

[a] KIP proteins (p21, p27) bind multiple cyclin-Cdk complexes and prevent activation or block kinase activity.
[b] INK proteins (p15 and p16) are specific for Cdk4/6 and cyclin D. They bind Cdk and prevent the binding of cyclin D.

B Regulation of cell cycle progression by Cdk–cyclin complexes

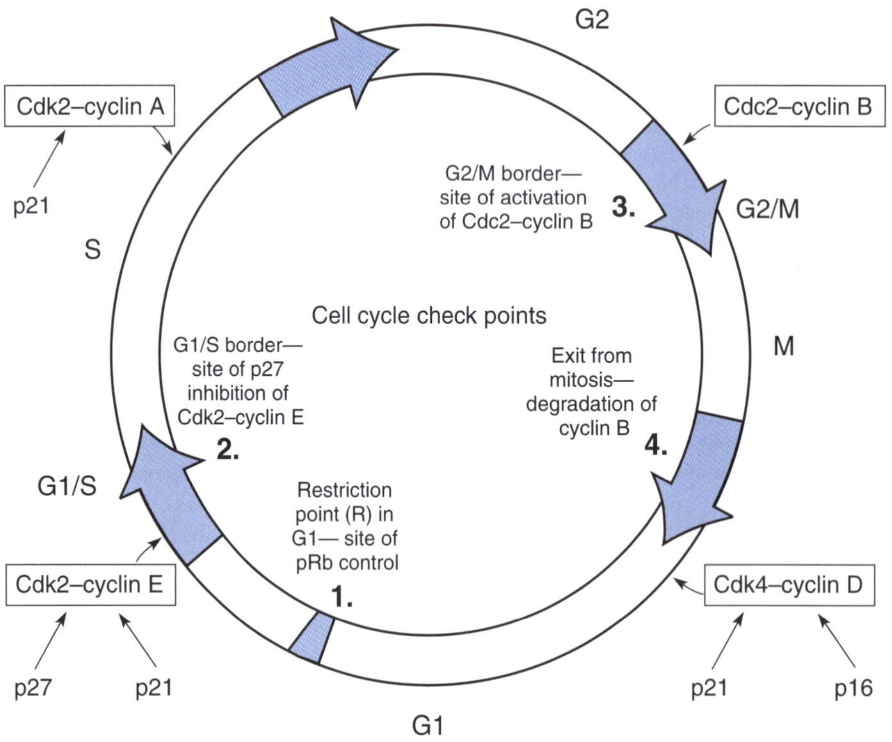

OVERVIEW

The processes by which a cell replicates its genetic material, equally divides it, and transfers it to two new daughter cells are the keys to the normal growth and development of any organism. In human cells, these processes take place during the well-regulated and ordered progression of the mammalian cell cycle. Control of the cell cycle ultimately determines whether a cell will continue to cycle, grow, and divide; will undergo differentiation; or will become quiescent and stop cycling. The mammalian cell cycle consists of four distinct phases: G1, S, G2, and M. The G1, G2, and S phases make up the time in the cell cycle between cell divisions; together they are referred to as the interphase portion of the cell cycle. Cells not undergoing cell division, such as neuronal cells, are removed from the cell cycle and exist in a phase called G0. If cells in G0 are stimulated to grow, they enter from G0 into the G1 phase, G1 being the start of the cycle. The functions of G1 phase proteins and their regulators are related to biochemical processes that control cell growth, differentiation, signal transduction pathways, aging, cancer, and programmed death. Cells in the G1 phase react to environmental stimuli (e.g., growth factors, inducers of differentiation, and even negative factors) until the restriction point (R point) in G1 is reached. Once the cell passes this point, it is committed to the S phase and dividing. During the S phase of the cell cycle, the total complement of DNA must be replicated faithfully, but only once. After the DNA has been completely replicated, the cells proceed to G2, during which time the chromatin begins to condense and form well-defined structures. A crucial point in the cell cycle occurs at the G2/M checkpoint, when the cells will decide whether to enter M, or mitosis. During the M phase, the duplicated chromosomes are equally distributed to two daughter cells, each of which can begin the cycle again.

Proteins Involved in Regulation of the Cell Cycle

Cell cycle progression is driven by biochemical reactions with multiple protein kinases, known as cyclin-dependent kinases (Cdk), being sequentially activated by the binding of proteins known as cyclins (Figure **A**). Once activated, each Cdk–cyclin protein complex phosphorylates specific key proteins that are required for the reactions that take place in each phase of the cell cycle. Orderly progression through the cell cycle depends on positive factors that drive the cycle forward, and negative factors that stop or arrest the cycle at a particular stage. The main positive factors are specific Cdks and specific cyclins that function at each stage of the cycle. Negative factors act by blocking the activity of the specific Cdks and are known as cyclin-dependent kinase inhibitors (CKI).

Regulation of Cell Cycle Progression by Cdk–cyclin Complexes

Cdks, although present in a cell, are not enzymatically active until they form a complex with a particular cyclin. Cyclins are a class of proteins that vary dramatically in level during the cell cycle. For example, the level of cyclin B increases continuously during the interphase, then declines precipitously during the M phase. These changes in the level of cyclin B are correlated with the activity of a specific Cdk, known as Cdc2. Cdc2 is active when cyclin B levels peak and becomes inactive as cyclin B levels suddenly decline. Thus, the phosphorylating activity of Cdc2 is modulated during the cell cycle by the availability of cyclin B. Active Cdk–cyclin complexes can be inactivated by several mechanisms, including cyclin degradation, alteration of the phosphorylation state, or interaction with CKI molecules. The interplay between the activation and deactivation of the Cdk activities at various steps of the cell cycle is the key to the normal progression and regulation of the cell cycle, with each phase of the cycle offering a site or sites for specific control (Figure **B**). The key positive regulators in the G1 phase are Cdk4 and cyclins of the D family, which form a complex capable of phosphorylating various proteins required to function in G1. One key protein subject to phosphorylation by Cdk4–cyclin D in G1 is the retinoblastoma protein (pRb). This protein, coded for a gene that maps to chromosome 13 at 13q14, is mutated in some human cancers, in particular retinoblastoma in children. When pRb is not phosphorylated, it restricts further cell growth by blocking the progression of the cell and preventing it from moving past the R point (*checkpoint 1*) in G1 and into the S phase. If, however, environmental conditions are such that the cell is stimulated to grow and divide, pRb becomes phosphorylated, and the cell continues past the R point and on to the S phase (see Chapter 23). Another important control that regulates the progression of a cell from G1 to S uses the CKI p27 which binds to the Cdk2–cyclin E complex and inactivates it. With p27 bound to the Cdk2–cyclin E complex, the cells are unable to proceed into the S phase and remain arrested in G1 (*checkpoint 2*). Growth stimulation factors can result in the degradation of p27, activation of the Cdk2–cyclin E complex, and transition to the S phase. The entrance into the S phase from G1, and progression through S to G2, is also dependent on the functioning of specific Cdk–cyclin complexes. Cdk2 initially binds cyclin E as the cells proceed into the S phase. Just before measurable DNA synthesis occurs, cyclin A binds and activates Cdk2, and phosphorylation of the proteins essential for DNA replication occurs. The critical proteins involved in the G2/M checkpoint (*checkpoint 3*) include Cdc2 and cyclin B. The activation of the Cdc2–cyclin B complex involves a series of well-controlled phosphorylation-dephosphorylation steps and is essential for the entrance into M. Once the Cdc2–cyclin B complex has been activated, mitosis begins. The exit from mitosis and the entrance to G1 occurs with the degradation of cyclin B and inactivation of the Cdc2–cyclin B complex (*checkpoint 4*).

Clinical Significance

Loss of control of the cell cycle leads to loss of control of normal growth and often results in abnormal cell growth (tumor cells or developmental defects) or programmed death of a cell (apoptosis). Many tumor suppressor proteins, such as pRb and p53, function during the cell cycle to regulate cell growth and division, thus preventing cancer development. Arrest in the G1 phase is of particular importance when DNA damage has occurred and must be repaired before DNA replication.

For more information see Hoffee PA: *IMS: Medical Molecular Genetics.* Fence Creek, pp. 92–99.

6 DNA Replication

A DNA polymerases involved in DNA replication

DNA polymerase	*Size (catalytic subunit) (kilodaltons)*	*Location*	*Function in the cell*
a	160–185	Nucleus	Lagging strand replication
b	40	Nucleus	DNA repair
g	125	Mitochondria	Replication of mitochondrial DNA
d	125	Nucleus	Leading and lagging strand replication
e	210–230	Nucleus	DNA repair (?)

B Multiple replication origins along the chromosome

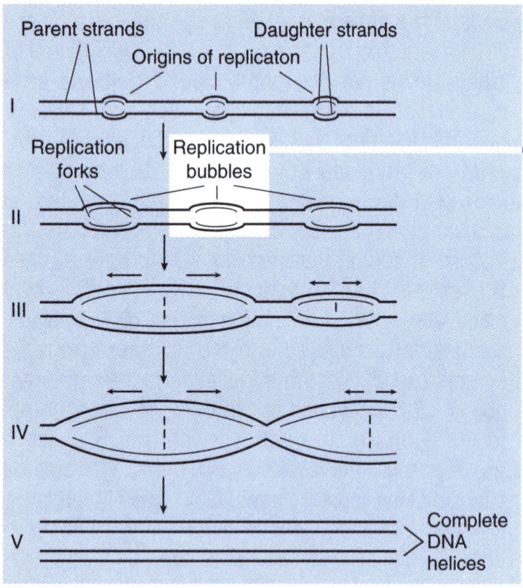

C Replication is discontinuous

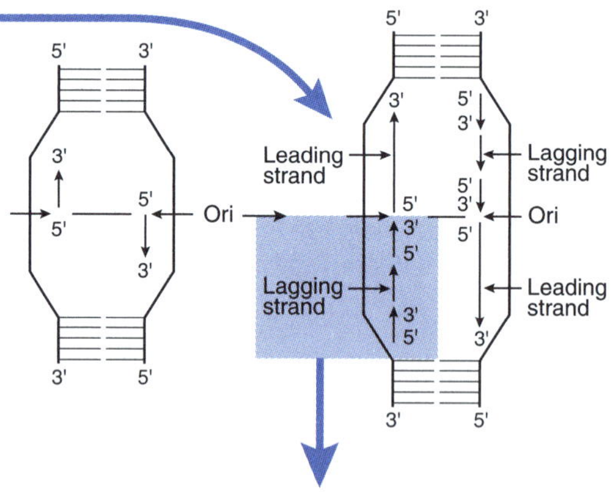

D Model for replication of DNA

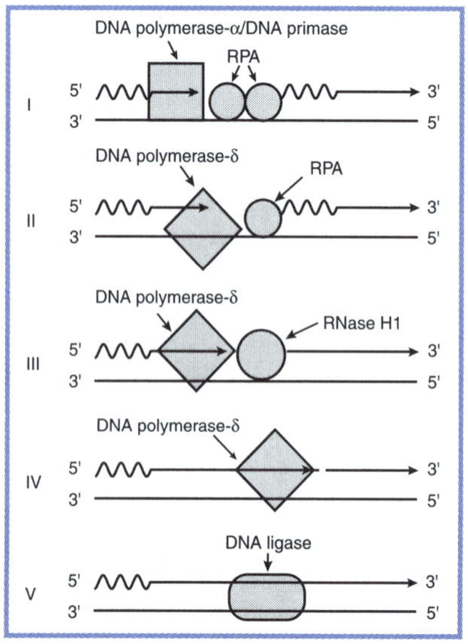

E Proteins involved in DNA replication

Protein	*Function*
DNA helicase	Unwinds DNA and breaks hydrogen bonds
Single-stranded DNA–binding protein (RPA)	Binds single-stranded DNA to prevent hydrogen bonding
Proliferating cell nuclear antigen (PCNA)	Stimulates DNA polymerase-δ activity
DNA polymerase-δ	Leading and lagging strand DNA replication and 3'–5' exonuclease proofreading
DNA polymerase-α/DNA primase complex	Synthesis of RNA primers and lagging strand synthesis
DNA ligase	Seals 3' terminal hydroxyl and 5' terminal phosphate group of adjacent nucleotides in DNA
Ribonuclease H1 (Rnase H1)	Removes RNA from a RNA–DNA hybrid
DNA topoisomerase	Relaxes DNA by breaking and resealing phosphodiester bonds

Note: RPA = replication protein A

OVERVIEW

Cells transmit genetic material to their progeny by replicating their entire genome before cell division. The 46 chromosomes of each human cell contain more than 6 billion bp of DNA, which must be completely replicated during the S phase of the cell cycle, or in approximately 7–8 hours, without mistakes and without rereplicating or missing parts of the genome. Each chromosome is a linear double-stranded DNA molecule with the two polynucleotide strands in opposite polarity held together by hydrogen bonding between the A-T and G-C base pairs. During DNA replication, the two strands separate, with each strand serving as a template for the synthesis of a new daughter strand. This mode of replication is referred to as semiconservative replication. After the first round of DNA synthesis and cell division, each of the two daughter cells contains one old strand of DNA and one newly synthesized strand of DNA. Because of the specificity of the hydrogen bonding, wherein A pairs with T and G with C, the sequence of bases in the template strand will dictate the sequence of bases in the newly synthesized strands. In this way, the nucleotide sequence of the DNA molecule is maintained in both daughter cells and is identical to that in the original parental molecule.

DNA Polymerases Involved in DNA Replication

DNA polymerases catalyze the formation of the phosphodiester bonds between adjacent deoxyribonucleotides in the DNA chain, using the template strand of DNA and the hydrogen-bonding properties of the purines (A,G) and the pyrimidines (C,T) to determine which deoxyribonucleotide will be added to the 3′ hydroxyl of the growing chain. All DNA polymerases catalyze the synthesis of DNA only in the 5′ to 3′ direction. Human cells contain five known DNA polymerases, referred to as DNA polymerase α, β, γ, δ, and ε. Each DNA polymerase serves a different function in the cell, as indicated in Figure **A**.

Multiple Replication Origins along the Chromosome

DNA replication initiates on each human chromosome at specific positions, referred to as the origins of replication (oris). Multiple oris occur every 150–200 kbp on each chromosome, with some 30,000 initiation sites localized over the entire human genome (Figure **B**). The presence of many oris per chromosome results in multiple sections of the genome being replicated simultaneously. Each small replicating unit, or replicon, has its own ori site where DNA replication begins. Once initiated, the replication complex proceeds bidirectionally on the chromosome until each replication bubble, or replicon, comes in contact with the next one. In this way, the entire genome can be completely replicated during a single S phase.

DNA Replication Is Discontinuous

Consider a short segment of DNA containing an ori site with two strands of the DNA, one strand running in a 5′ to 3′ direction and the opposite strand running in a 3′ to 5′ direction (Figure **C**). As the two strands unwind and come apart, DNA synthesis initiates at the ori and proceeds down the two strands. However, because of the properties of the DNA polymerases, synthesis can proceed from the ori only in a 5′ to 3′ direction. One strand, therefore, is synthesized in a 5′ to 3′ direction, and the opposite strand is also synthesized in a 5′ to 3′ direction. Notice, however, that the two forks move away from each other. Since there is no known DNA polymerase that can function in a 3′ to 5′ direction, DNA cannot be synthesized in the 3′ to 5′ direction. The cell resolves this problem by synthesizing short fragments of DNA, known as Okazaki fragments, which are 200 nucleotides in length and synthesized in a 5′ to 3′ direction. The resulting fragments are then joined together by an enzyme called DNA ligase to give one continuous strand of DNA. The strand of DNA that is synthesized continuously from the 5′ to 3′ direction is known as the leading strand of DNA synthesis. The strand of DNA that is synthesized 5′ to 3′ in short pieces, or discontinuously, is called the lagging strand of DNA synthesis.

Model for Replication of DNA

During DNA replication of the lagging strand, the two DNA strands are pulled apart by DNA helicase, exposing single-stranded regions on the DNA (Figures **D** and **E**). RPA, a single-stranded DNA-binding protein, binds the DNA to keep the strands apart. The DNA polymerase-α–DNA primase complex then binds to the DNA and begins the synthesis of an RNA primer. Using the RNA primer, DNA polymerase-α synthesizes a small stretch of DNA by the addition of deoxyribonucleotides to the 3′ terminus of the primer. The DNA polymerase-α complex is then dissociated from the DNA and replaced with DNA polymerase-δ, which continues replication of the lagging strand. After two Okazaki fragments have been replicated on the lagging strand, they each still contain a small RNA primer. Before DNA replication can be completed, the RNA primer must be removed from the DNA, a step carried out by RNase H1. Once the RNA has been removed from the DNA, the gap that remains is filled in by DNA polymerase-δ, using the opposite DNA strand as a template. Finally, the nick remaining between the two adjacent Okazaki fragments is closed by the action of DNA ligase to give a large, continuous, newly synthesized DNA strand.

Clinical Significance

The occurrence of a single base change in the DNA sequence during replication can result in a mutation that leads to a defective protein and human disease. Thus, fidelity of DNA replication over the entire genome is essential to prevent mistakes in the DNA sequence from occurring. Additionally, mutations that lead to defects in DNA helicase activity, an activity essential for DNA replication, have been shown to result in Werner syndrome, which is characterized by premature aging, and Bloom syndrome, which is characterized by genetic instability and the early onset of neoplasia.

For more information see Hoffee PA: *IMS: Medical Molecular Genetics*. Fence Creek, pp. 23–32.

7 Mitosis

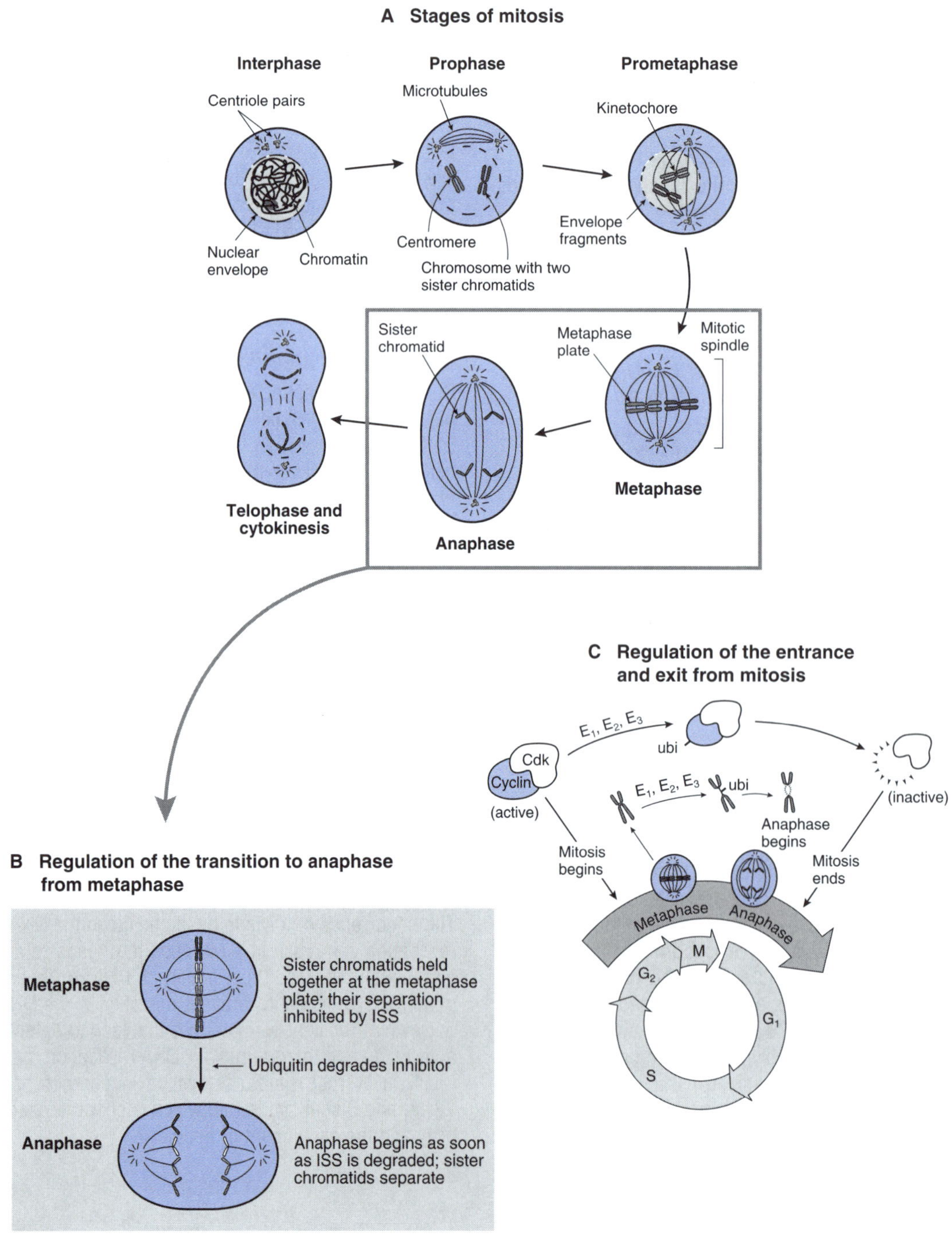
A Stages of mitosis
Interphase
Centriole pairs
Nuclear envelope
Chromatin
Prophase
Microtubules
Centromere
Chromosome with two sister chromatids
Prometaphase
Kinetochore
Envelope fragments
Sister chromatid
Metaphase plate
Mitotic spindle
Metaphase
Anaphase
Telophase and cytokinesis
B Regulation of the transition to anaphase from metaphase
Metaphase
Sister chromatids held together at the metaphase plate; their separation inhibited by ISS
Ubiquitin degrades inhibitor
Anaphase
Anaphase begins as soon as ISS is degraded; sister chromatids separate
C Regulation of the entrance and exit from mitosis
E_1, E_2, E_3
ubi
Cdk
Cyclin
(active)
(inactive)
E_1, E_2, E_3
ubi
Anaphase begins
Mitosis begins
Mitosis ends
Metaphase
Anaphase
M
G_2
G_1
S

OVERVIEW

Mitosis, or the M phase, is the portion of the human cell cycle in which the condensed chromosomes are equally distributed to two new daughter cells. Important transition stages are now recognized in the mitotic phase of the cell cycle. One is the sudden activation of the Cdc2–cyclin B complex by dephosphorylation, which occurs at the G2/M border and results in the phosphorylation of many proteins required for mitosis. Mitosis involves several significant cellular changes, which include the condensation of the chromosomes, dissolving of the nuclear membrane, mitotic spindle formation, and changes in cell surface proteins. The proteins involved in these reactions—such as lamins, histones, actins, and cytoskeleton proteins, to name just a few—are subject to phosphorylation by the activated Cdc2–cyclin B complex, preparing them for their role in mitosis.

Stages of Mitosis

Mitosis occurs in five distinct phases: prophase, prometaphase, metaphase, anaphase, and telophase (Figure **A**). *Prophase* is the stage at which the chromosomes have condensed to the point where they are now visible in the light microscope as thin threads. The cells have just passed through an S phase, and each chromosome has been duplicated once and consists of two sister chromatids connected by a centromere, with each chromatid containing a DNA–protein complex known as a kinetochore. Microtubules attach to each of the kinetochores and link the centromere of each chromosome to the spindle poles. Cytoplasmic microtubules are broken down during prophase and reorganized into the mitotic spindle. *Prometaphase* starts with the fragmentation of the nuclear envelope, followed by the movement of the mitotic spindle into the area and attachment of some of the microtubules to the kinetochores on each chromosome. At this point the chromosomes begin to move to the center of the cell, where they will become aligned at the metaphase plate. At *metaphase*, all of the chromosomes are aligned at the metaphase plate, at which point the process is arrested for 20 minutes. When the appropriate signal is given, *anaphase* begins, and the sister chromatids are pulled apart, one chromatid from each chromosome moving to opposite poles of the cell. In this way, one chromatid from each replicated chromosome is passed to each daughter cell. The final phase, *telophase*, finds the chromatids present at the two poles. At this point, the processes that occurred at the beginning of mitosis are reversed, with decondensation of the chromosome, disassembly of the spindle apparatus, and re-formation of the nuclear envelopes around the daughter chromosomes. Cytokinesis then occurs, resulting in two new daughter cells that are prepared to begin the cycle again.

Regulation of the Transition to Anaphase from Metaphase

Recent evidence has implicated a critical control point occurring in mitosis between metaphase and anaphase (Figure **B**). A protein, an inhibitor of sister chromatid separation (ISS), binds to the chromatids and keeps the two sister chromatids together until all of the chromosomes are properly aligned on the metaphase plate. At a crucial point, the protein ubiquitin directs the degradation of the ISS protein, and anaphase begins with the sister chromatids moving to each pole. This mechanism helps assure that anaphase does not begin until all of the chromosomes are properly aligned, and it prevents the occurrence of daughter cells with the improper number of chromosomes.

Regulation of the Entrance into and the Exit from Mitosis

Three important transition stages are now recognized in the mitotic phase of the cell cycle. They are key to a properly ordered entrance into and exit from mitosis, with each daughter cell receiving an exact copy of the parental genome. These stages are (1) the transition from G2 to M concurrent with the activation of the Cdc2–cyclin B complex; (2) the M-phase checkpoint that occurs in metaphase, the point that regulates the timing of the separation of the chromatids and the initiation of anaphase; and (3) at the onset of anaphase, the sudden proteolytic destruction of cyclin B with the simultaneous inactivation of Cdc2, allowing the cell to prepare for exit from M into a new G1 phase. The latter two checkpoints entail the sudden degradation of two specific proteins, the ISS protein and cyclin B, and are regulated by a common pathway that involves the action of a proteolytic complex known as the ubiquitin–proteasome pathway (Figure **C**). The ubiquitin–proteasome pathway, a complex multistep pathway involving three enzymes—E1, E2, and E3—results in the covalent attachment of polyubiquitin chains to proteins subject to degradation. The polyubiquitinated protein is then degraded by a proteasome complex to small peptides. The advantages of this destruction method to control the levels of a particular protein are the suddenness of the reaction, resulting in instant control, and the irreversibility of the action. Only with new protein synthesis can a cell once again accumulate significant levels of the target protein, giving the cell additional levels of control over the expression of certain proteins.

Clinical Significance

Accurate control of the events in mitosis is necessary to ensure that after cell division each new daughter cell contains a complete complement of chromosomes identical to the parental cell. Multiple control points involving phosphorylation of proteins and destruction of proteins are utilized to this end. Any deviation from the equal distribution of chromosomes to the daughter cells gives rise to cells with an increased or decreased number of chromosomes.

For more information see Hoffee PA: *IMS: Medical Molecular Genetics*. Fence Creek, pp. 96–99.

8 Meiosis

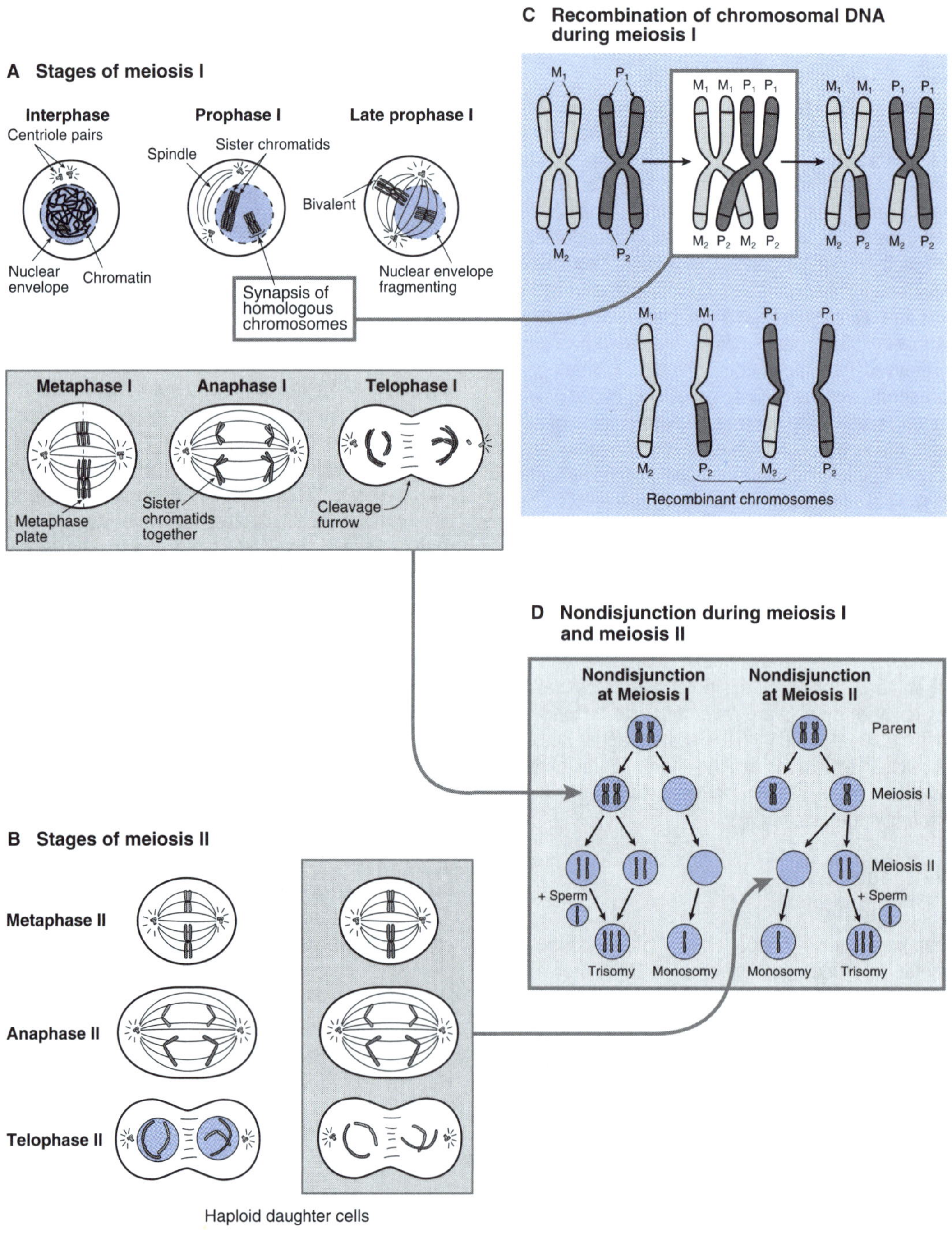
A Stages of meiosis I
Interphase
Centriole pairs
Nuclear envelope
Chromatin
Prophase I
Spindle
Sister chromatids
Synapsis of homologous chromosomes
Late prophase I
Bivalent
Nuclear envelope fragmenting
Metaphase I
Metaphase plate
Anaphase I
Sister chromatids together
Telophase I
Cleavage furrow
B Stages of meiosis II
Metaphase II
Anaphase II
Telophase II
Haploid daughter cells
C Recombination of chromosomal DNA during meiosis I
M_1 P_1 M_2 P_2
Recombinant chromosomes
D Nondisjunction during meiosis I and meiosis II
Nondisjunction at Meiosis I
Nondisjunction at Meiosis II
Parent
Meiosis I
Meiosis II
+ Sperm
+ Sperm
Trisomy
Monosomy
Monosomy
Trisomy

OVERVIEW

Preservation of the diploid chromosome number in human cells is accomplished during the process of meiosis, in which gametes are produced with a haploid number of chromosomes. If meiosis did not occur, the process of fertilization would result in a doubling of the chromosome number in a cell at each generation. Meiosis, which occurs only in the germ cells, maintains the diploid chromosome number by having a single genome duplication, which is then followed by two successive division cycles: meiosis I and meiosis II. Meiosis plays a major role in creating the genetic diversity we see in every child born of the same two parents. This diversity is accomplished by two means. First, various combinations of chromosomes (i.e., some of maternal origin and some of paternal origin) are randomly distributed to the gametes and passed on to the children. More than 1 million possible combinations of alleles can be produced by a single individual. Second, during meiosis I, recombination between synapsed (paired) homologous chromosomes occurs in prophase I, allowing new combinations of genes to be transferred to the gametes.

Stages of Meiosis I

Meiosis I is unique in that duplicated chromosomes become linked to each other in homologous pairs (i.e., the maternal and paternal chromosomes pair). This pairing, termed synapsis, occurs during prophase I of meiosis I and involves the alignment of two pairs of sister chromatids to give a bivalent made up of four chromatids (Figures **A** and **C**). The chromatids are now in position to allow crossing over between homologous chromosomes, forming chiasma (site of physical exchange of genetic material). The chiasma physically holds the two homologues together as the chromosome pair moves to the metaphase plate. At metaphase I, all of the chromosome pairs, or bivalents, become aligned at the metaphase plate with microtubules attached to their kinetochores. Anaphase I begins when the two homologous chromosomes are pulled apart and begin to move to each pole. Note that in meiosis I, it is the chromosomes, each of which is made up of two sister chromatids, that separate and move to the poles, as opposed to meiosis II and mitosis, in which sister chromatids separate at anaphase and move to opposite poles. At telophase, one chromosome of each of the 23 pairs is found in the two daughter cells. The cells at this stage have 23 chromosomes, random with respect to their parental origin, and 46 chromatids.

Stages of Meiosis II

Meiosis II, which resembles mitosis, begins shortly after the completion of meiosis I. The 23 chromosomes are present in the cells as duplicated DNA molecules, each of which contains two sister chromatids. During prophase II, the microtubules attach to the kinetochores and begin to move the chromosomes to the center of the cell. At metaphase II, the 23 chromosomes, some of which are recombinant chromosomes, are aligned on the metaphase plate (Figure **B**). At the beginning of anaphase II, the centromeres divide, and sister chromatids move to opposite poles. The final stage of telophase II and cytokinesis produces four haploid cells with 23 chromosomes and 23 chromatids.

Recombination of Chromosomal DNA during Meiosis I

Recombination of chromosomal DNA results from a physical breakage on the chromatids of two homologous chromosomes (one maternal chromosome and one paternal chromosome) with a reciprocal exchange of genetic material. The resulting chromosomes now contain genetic material from both parents and are referred to as recombinant chromosomes (Figure **C**). The frequency of recombination occurring between two genes on a single chromosome is dependent on the distance between the two genes (i.e., genes that are very close together will show a low frequency of recombination, whereas two genes that are far apart will have an increased frequency of recombination). The ability to identify recombinant progeny and calculate the recombination frequency between two genes allows the creation of genetic maps, which indicate the location of genes along a chromosome and their relationship to each other in terms of the distance between them (see Chapter 48).

Nondisjunction during Meiosis I and Meiosis II

Aneuploidy is the term used to describe a cell in which an individual chromosome of one pair is found missing (monosomy) or a chromosome is present as an extra copy (trisomy). Monosomy and trisomy arise by the process of nondisjunction, wherein the two sister chromatids or two homologous chromosomes fail to separate during anaphase (Figure **D**). Both meiosis I and meiosis II can give rise to nondisjunction. If nondisjunction occurs during meiosis I or II, one of the gametes produced will be missing one chromosome, while the other will have an additional copy of the chromosome. Fertilization by a normal sperm then gives rise to either trisomy or monosomy in the offspring. Monosomy and trisomy also can result from anaphase lag, or the delayed movement of a single chromosome toward the poles during anaphase. Both nondisjunction and anaphase lag have the same consequence, producing one cell that is missing a chromosome and one cell with an additional copy of a chromosome. Human cells have evolved a checkpoint to try to reduce the incidence of anaphase lag by delaying the initiation of chromatid or chromosome separation until all of the chromosomes are properly aligned on the metaphase plate (see Chapter 7).

Clinical Significance

Many chromosome abnormalities result from mistakes that occur during meiosis. Nondisjunction can occur during either meiosis I or meiosis II, giving rise to monosomies or trisomies. Monosomies, unless they involve a sex chromosome, are almost always lethal, as are most trisomies. However, three trisomies found in humans are compatible with life. They include trisomy 21 (Down syndrome), trisomy 13 (Patau syndrome), and trisomy 18 (Edwards syndrome). The frequency of meiotic nondisjunction increases with increasing maternal age, which is reflected in the increased risk in older women of having a child with Down syndrome (trisomy 21). In addition, in meiosis I, mistakes occurring during the recombination process can result in chromosomal deletions, inversions, translocations, and duplications.

For more information see Hoffee PA: *IMS: Medical Molecular Genetics.* Fence Creek, pp. 99–102.

9 Transcription

A RNA polymerases involved in transcription

RNA polymerase	*Location*	*RNAs synthesized*
I	Nucleus	rRNAs—28S, 18S, 5.8S
II	Nucleus	precursor mRNA, snRNAs
III	Nucleus	tRNAs, 5S rRNA
Mitochondrial	Mitochondria	Mitochondrial RNA

B Important aspects of a typical human gene

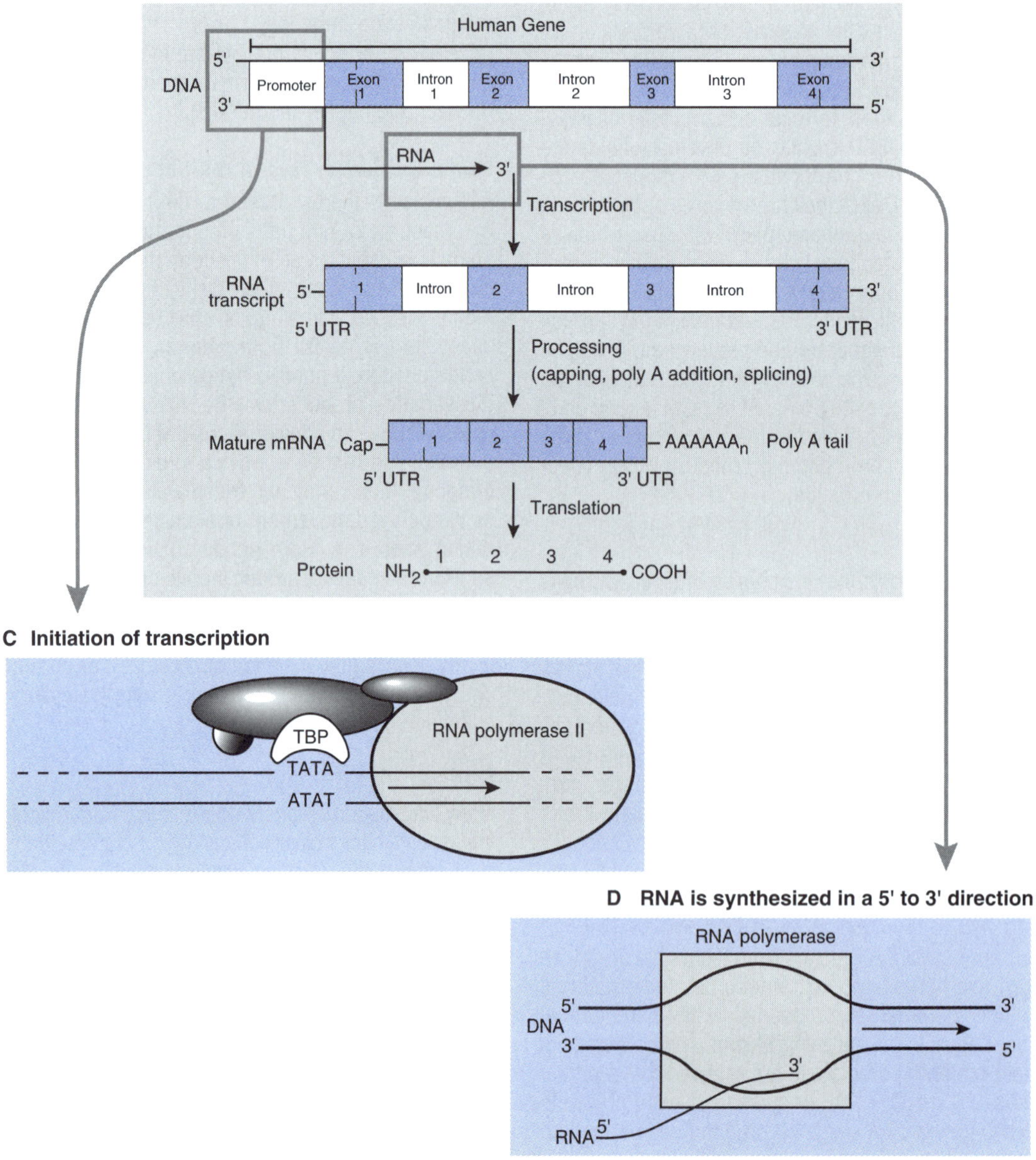

OVERVIEW

The biologic information stored in a DNA molecule must find a way to be expressed in a cell. This expression is accomplished by transcribing the information in DNA into different forms of RNA, a process carried out by RNA polymerase. RNA polymerase catalyzes the formation of phosphodiester bonds between adjacent ribonucleotides in the growing RNA chain, using one strand of the DNA as a template. In the nucleus of human cells there are three different types of RNA polymerases, referred to as RNA polymerase I, II, and III (Figure **A**). Additionally, mitochondria have their own RNA polymerase. RNA polymerase I transcribes the genes that code for three of the four ribosomal RNAs (28S, 18S, and 5.8S); RNA polymerase III transcribes the genes that code for tRNAs and the 5S rRNAs; and RNA polymerase II transcribes mRNA starting from a TATA box within a promoter region (see below) and also synthesizes some of the snRNAs. Each RNA polymerase is a complex molecule of high molecular weight and is made up of multiple subunits.

Important Aspects of a Typical Human Gene

A typical gene found in the nuclear DNA of a human cell has certain features that are common to all human genes that code for proteins (Figure **B**). Each gene is made up of a particular sequence of base pairs with the size of different genes ranging from tens to millions of kbp of DNA. The sequence of base pairs is specific for each gene and determines the specificity of the product coded for by that gene. However, not all of the bases present in a gene are expressed in the final product. Human genes are often split into (1) sequences that remain in the final mature mRNA and code for information (exons) and (2) sequences that are removed from the primary RNA transcript during processing (introns). In addition to coding sequence information, exons contain other sequences that are necessary for the functioning of a mRNA. Thus, an exon is defined as a sequence in a primary RNA transcript that is conserved during the processing of the transcript into a mature mRNA. The initial RNA transcript contains both intron and exon sequences and must be processed to the mature mRNA in a series of steps. These steps include the addition of a Cap structure at the 5′ end of the mRNA, the addition of a poly A tail at the 3′ end of the mRNA, and finally the removal of the intron sequences and the splicing together of the exon sequences to produce the final mRNA product. One additional feature of the mature mRNA molecule is the presence of untranslated regions (UTRs) found at both the 3′ and the 5′ ends of the molecule. These are referred to as 3′UTR or 5′UTR and represent sequences in the exons that remain in the mRNA but will not be translated into protein. These regions contain regulatory signals that are necessary both for processing of the mRNA and for its translation into protein. In human genes, the amount of DNA represented as intron sequences far exceeds that present in exon sequences.

Initiation of Transcription

Base sequences, such as the TATA sequence and the initiator (Inr), that are important for the initiation of transcription are present at the 5′ end of most human genes and represent promoter sequences (Figure **C**). The function of the promoter is to set the location and the direction of transcription on a DNA template. RNA polymerase II must be recruited to the promoter site, a task mediated by a large number of proteins called transcription factors (TFs). TFs can bind to sequences in the promoter or can bind to one another in many different arrays to instruct RNA polymerase II whether to transcribe a particular gene (see Chapters 17–20). Different cells in the body contain identical DNA sequences, but they may express different genes at different times and in response to different signals. Controlling the variation of expression of genes is the function of the interaction of various TFs within different cells. Some basic steps occur in the initiation of transcription of all genes, one of which is the binding to the TATA sequence in a promoter by a protein known as the TATA-binding protein (TBP). Once the TBP is bound to the TATA sequence in the promoter, additional TFs then bind either to the TBP or directly to the DNA, resulting in a protein–DNA complex, which now recruits RNA polymerase II to the promoter site. RNA polymerase II binds to the DNA about 25 bp from the TATA box and initiates transcription in a specific direction (5′ to 3′). Only the base sequence of one of the two DNA strands, the template strand, is copied; the directionality of the transcription process is determined by the presence of additional sequences in the promoter, such as the Inr.

RNA is Synthesized in a 5′ to 3′ Direction

Before transcription by RNA polymerase II can begin, the two DNA strands must be pulled apart by the action of DNA helicases, an activity associated with some TFs. The separation of the two DNA strands to form single-stranded regions of DNA is necessary for the transcription process to begin and continue along a DNA molecule (Figure **D**). Once the DNA strands have been pulled apart, only one strand, that running in a 3′ to 5′ direction, serves as the template for RNA synthesis. RNA polymerase II catalyzes the formation of a phosphodiester bond by attaching a 5′ phosphate of the incoming ribonucleotide to the 3′ hydroxyl of the growing RNA chain. Multiple RNA molecules can be synthesized from a single DNA molecule with the binding of additional RNA polymerase molecules to the promoter sequence.

Clinical Significance

The sequences of bases found in DNA must be accurately transcribed into a mRNA to produce the proper protein structure in the right cell at the right time during cell growth or differentiation. Precise control of the initiation of transcription by multiple transcription factors and RNA polymerase II is essential for this process. Any deviation in the quantity or the quality of a mRNA in a particular cell can lead to an inherited human disorder. Many examples of these defects as the causes of inherited human disorders are known and will be discussed later.

For more information see Hoffee PA: *IMS: Medical Molecular Genetics.* Fence Creek, pp. 38–40.

10 RNA Processing

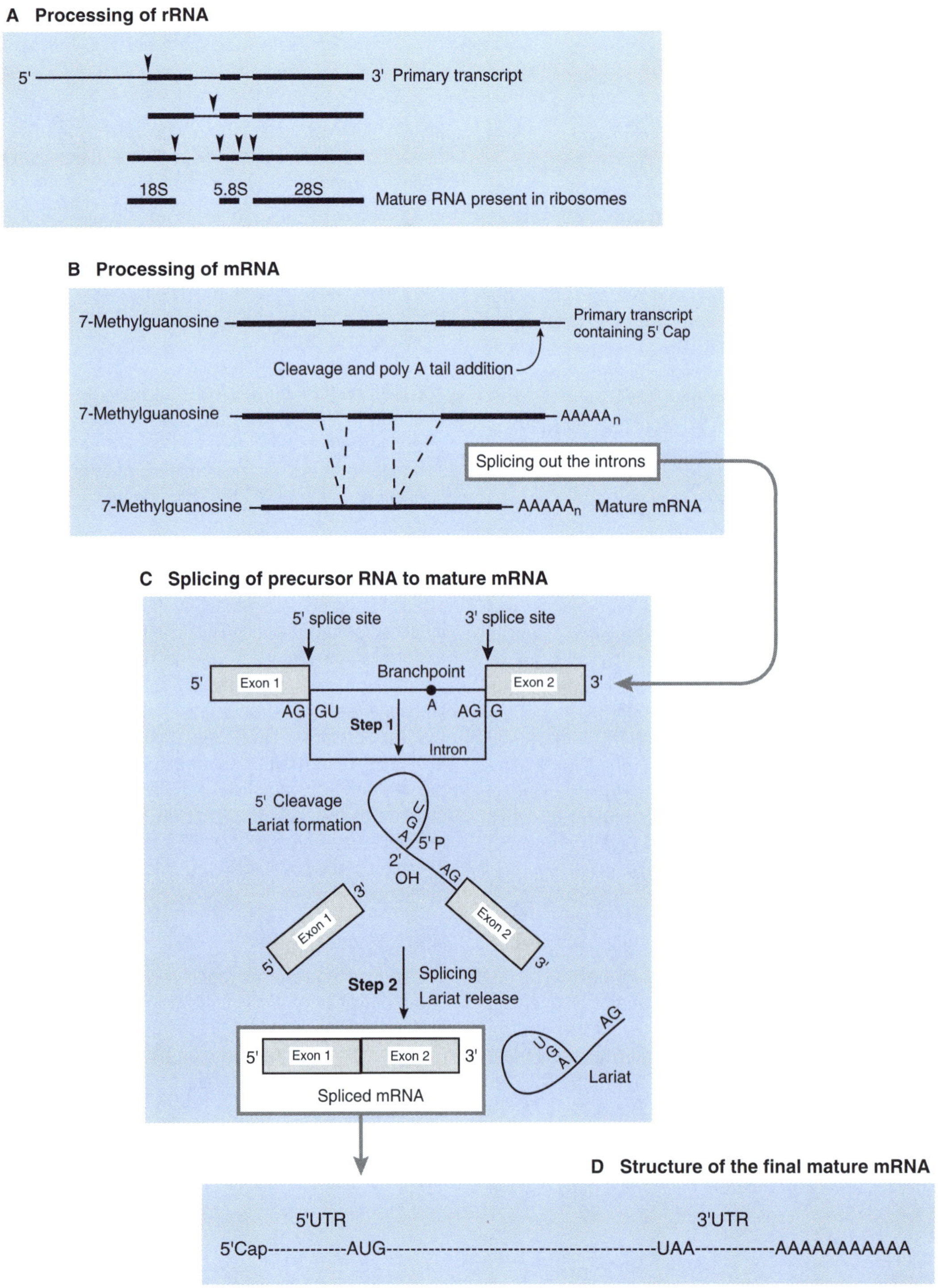

OVERVIEW

The primary transcript of RNA made directly from the DNA template is not the final biologic form of the RNA. RNAs, including rRNA, tRNA, and mRNA, must be processed before they attain the form that will function in protein synthesis.

Processing of rRNA and tRNA

rRNA is synthesized and processed in a small body in the nucleus known as the nucleolus. Processing of rRNA includes methylation of the RNA and removal of portions of the primary transcript by RNases (Figure **A**). A single RNA transcript produced by RNA polymerase I carries the information for three of the mature rRNAs found in ribosomes: the 18S, the 28S, and the 5.8S rRNAs. Processing of the transcript occurs with cleavage of the primary transcript by RNase at precise sequences. The initial cleavage gives a RNA product, which is further cleaved to give the 18S, the 5.8S, and the 28S mature RNAs. Molecules of tRNA are processed in a similar way, with the primary transcript being cleaved to smaller forms by RNases.

Processing of mRNA

The major RNA processing in the cell occurs with mRNA. mRNA is synthesized by RNA polymerase II as a large primary transcript known as hnRNA, which contains both intron and exon sequences. Once the transcript has been made by RNA polymerase II, three independent processing events occur before the mRNA is competent to take part in protein synthesis (Figure **B**). The first step in the processing of hnRNA is capping, or the attachment to the 5′ end of the RNA of a nucleoside called 7-methylguanosine (7-methyl G). 7-Methyl G is attached through a 5′ to 5′ phosphodiester bond between the ribose on the 7-methyl G and the first ribose on the RNA. This bond is unique, since all the other phosphodiester bonds in RNA and DNA are 3′ to 5′. The addition of 7-methyl G, or the Cap structure, makes the resulting RNA molecule more resistant to exonucleases in the cell and increases the stability of the RNA. The second step in the processing of hnRNA occurs at the 3′ end of the molecule in an area referred to as the 3′UTR. Within this 3′UTR is the sequence AAUAAA, known as the poly A addition signal. This sequence, located at the 3′ end of each primary RNA transcript, is recognized by a nuclease that cleaves the RNA, 11–30 nucleotides past this site. Following the cleavage, an enzyme called poly A polymerase recognizes the poly A addition signal and adds residues of adenosine to the end of the mRNA to give molecules of RNA that have 50–200 adenine residues (poly A tails) at their 3′ ends. The final step in the processing of hnRNA, which now carries a Cap at the 5′ end and a poly A tail at the 3′ end, is the removal of the intron sequences and the joining together of the exon sequences. This process, called splicing, occurs on small ribonuclear bodies known as spliceosomes.

Splicing of Precursor RNA to Mature mRNA

Spliceosomes function by recognizing the splice donor and splice acceptor sequences that occur at the beginning and end of each intron (Figure **C**). The splice donor site, GU, begins each intron, and the splice acceptor site, AG, ends each intron. In addition, within each intron sequence is a site known as a branch point. Within the branch point is an adenosine residue, which, in the presence of a spliceosome, combines with the G of the donor splice site. This reaction results in the release of the first exon and the formation of an intermediate structure known as the lariat. In the next step, exon 1 combines with the acceptor exon 2 to form a continuous mRNA containing only the exon sequences, and the lariat is released. The RNA components of spliceosomes are known as snRNAs. The SnRNP particle or spliceosome contains several snRNA molecules referred to as U-1, U-2, and so on, up to U-6. (U refers to the high content of uridine found in the snRNA). The removal of introns and the splicing of exons must be an extremely precise process, considering that most genes are made up of multiple exons and introns, and the recognition sequences GU and AG can occur in many places in the RNA, roughly every 20 bp. Thus, there are additional sequences 5′ and 3′ to the GU and AG sequences, which play an important role in the splicing process.

Structure of the Final Mature mRNA

The final mature mRNA consists of an RNA molecule with (1) a 7-methyl G Cap on the 5′ end to give the mRNA stability; (2) a run of adenines or a poly A tail at the 3′ end, which also contributes to stability and facilitates the transport of mRNA from the nucleus to the cytoplasm; (3) exon sequences both at the 5′ end and the 3′ end of the molecule, which are not translated into protein (UTR); and (4) exon sequences that are translated into protein (Figure **D**). At the beginning of the first translated exon there is an AUG sequence, which signals protein synthesis to begin. Toward the end of the last translated exon is a second signal (UAA, UAG, or UGA), which signals protein synthesis to terminate.

Clinical Significance

Even though the process of splicing is extremely accurate, when mistakes occur they can have disastrous effects on the protein product produced and generally result in a mutant phenotype. An example of splicing mutations is seen in the processing of the RNA for dystrophin, the defective protein in Duchenne muscular dystrophy. The dystrophin gene is made up of 79 exons and 78 introns spread over more than 2.5 million bp of DNA. The 79 exons are spliced together to give a mRNA of 14 kbp. When inaccurate splicing occurs, the resulting mRNA is missing all or part of an exon sequence, resulting in a totally inactive protein and a child with Duchenne muscular dystrophy. On the other hand, splicing can give a cell flexibility in using the same DNA sequence to make different proteins. For example, if a gene is made up of 20 exons, the cell may determine which exon sequences will be retained in the mature mRNA and thus be coding sequences for the final protein product. The proteins will be similar, but not identical, since they may share some exon sequences but differ in others. This process is referred to as differential or alternative splicing (see Chapter 24).

For more information see Hoffee PA: *IMS: Medical Molecular Genetics*. Fence Creek, pp. 40–43.

11 Translation

mRNA-Directed Protein Synthesis

A mRNA molecule

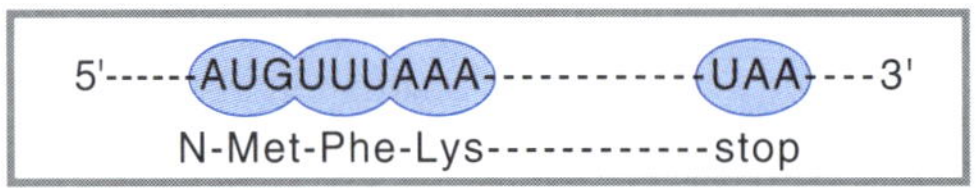

B The genetic code

First position	Second position U	Second position C	Second position A	Second position G	Third position
U	UUU Phe UUC Phe UUA Leu UUG Leu	UCU Ser UCC Ser UCA Ser UCG Ser	UAU Tyr UAC Tyr UAA Stop UAG Stop	UGU Cys UGC Cys UGA Stop UGG Trp	U C A G
C	CUU Leu CUC Leu CUA Leu CUG Leu	CCU Pro CCC Pro CCA Pro CCG Pro	CAU His CAC His CAA Gln CAG Gln	CGU Arg CGC Arg CGA Arg CGG Arg	U C A G
A	AUU Ile AUC Ile AUA Ile AUG Met/start	ACU Thr ACC Thr ACA Thr ACG Thr	AAU Asn AAC Asn AAA Lys AAG Lys	AGU Ser AGC Ser AGA Arg AGG Arg	U C A G
G	GUU Val GUC Val GUA Val GUG Val	GCU Ala GCC Ala GCA Ala GCG Ala	GAU Asp GAC Asp GAA Glu GAG Glu	GGU Gly GGC Gly GGA Gly GGG Gly	U C A G

D Interactions of codons and anticodons

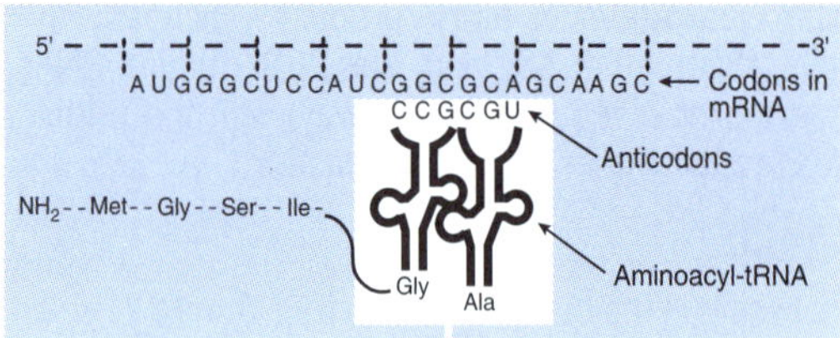

C tRNA: Translator of the genetic code

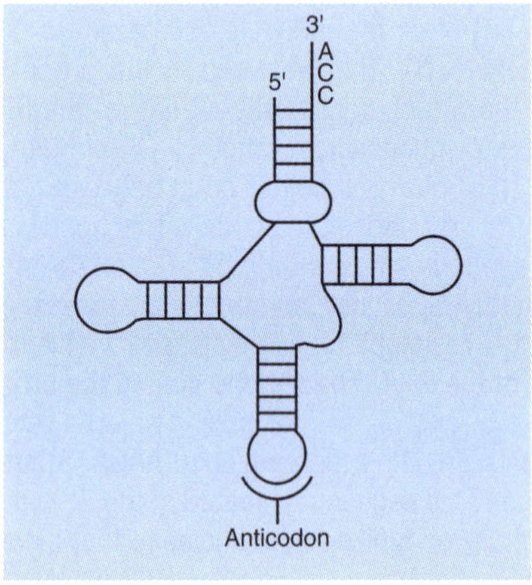

E Mutations: Changes in the DNA base sequence

Base Pair Substitution

Missense mutation

	Normal	Mutant
DNA	AAA	AGA
RNA	UUU	UCU
Protein	Phe	**Ser**

Nonsense mutation

	Normal	Mutant
DNA	AGC	ATC
RNA	UCG	UAG
Protein	Ser	**Stop**

Base Pair Addition or Deletion

Frameshift mutation

	Normal		Addition
DNA	TAC AAA TTT CAA AGC	DNA	TAC AAA **C**TT TCA AAGC
RNA	AUG UUU AAA GUU UCG	RNA	AUG UUU **G**AA AGU UUCG
Protein	Met-Phe-Lys - Val - Ser	Protein	Met - Phe - **Glu** - **Ser** - **Phe**

OVERVIEW

The mature mRNA molecules produced by the process of transcription and processing carry the information from the DNA nucleotide sequence to the ribosomes, where the information is translated by the ribosomes to produce a protein of a determined amino acid sequence. The mRNA, which contains in its structure instructions for the synthesis of proteins, is composed of a single strand of ribonucleotides held together by phosphodiester bonds (Figure **A**). At the 5′ end of mRNA is a signal for translation initiation, the AUG codon. Toward the 3′ end of the mRNA is a signal for translation termination, either UAA, UAG, or UGA. The sequence of bases in the mRNA between AUG and UAA determines the exact sequence of amino acids in the protein. The mRNA at this point is colinear with the protein to be produced, meaning that there are no breaks or interruptions in the mRNA sequence. Protein synthesis starts at AUG initiator codon and proceeds in a 5′ to 3′ direction, each ensuing triplet coding for the next amino acid in the protein until a termination codon is reached, ending protein synthesis.

The Genetic Code

The information in the mRNA sequence is read by way of the genetic code. The genetic code is made up of 64 codons, each containing three of the four bases: A, U, C, or G. All permutations of the four bases in groups of threes are represented (Figure **B**). Each of 61 triplets codes for a particular amino acid. For example, when the codon UUU appears in a mRNA molecule, a phenylalanine will be found at that position in the protein. Of the 64 possible codons, three of the codons, UAA, UAG, and UGA, do not code for an amino acid and are referred to as nonsense or termination codons. When one of these codons, UAA, UGA, or UAG, appears in the mRNA sequence, it is a signal to stop translation. The genetic code, with some minor exceptions, is universal, with the same codons always coding for the same amino acid. An exception to this general codon usage occurs in the mitochondria, where UGA codes for tryptophan, AUA for methionine, and CUA for threonine. More than one codon can code for the same amino acid. An example is found with the amino acid valine, which can be coded for by either GUU, GUC, GUA, or GUG. This is known as redundancy of the genetic code.

tRNA: Translator of the Genetic Code

tRNA is a critical component of protein synthesis. It is made in the nucleus as a precursor RNA and then processed to form the final product (Figure **C**). The nucleotide sequence CCA, which is added to the 3′ end of all tRNAs after transcription has occurred, represents the site at which specific amino acids attach. Located at the bottom loop of the tRNA molecule is the anticodon site, which is required to recognize a specific codon found in a mRNA.

Interactions of Codons and Anticodons

tRNA plays a crucial role in taking the information in the mRNA and turning it into a protein molecule. Every amino acid is recognized by a specific tRNA molecule and is attached to the CCA end of the tRNA (Figure **D**). The three-base sequence on the tRNA known as the anticodon is complementary to a specific codon found in the mRNA. The codon in the mRNA (i.e., GGC) is recognized by the anticodon of the tRNA, CCG, allowing the tRNA to attach to the mRNA by base complementarity. Because each tRNA carries a specific amino acid, the binding of the tRNA to the mRNA by base pairing between the codon and the anticodon provides the specificity for placing a particular amino acid in the growing polypeptide chain. In the example shown, a glycine is placed at the site of the GGC codon, and an alanine is placed at the site of the GCA codon. UAA, UAG, and UGA, the three termination codons, do not have a tRNA with a complementary anticodon; therefore, they are not recognized by a tRNA and no amino acid is added to the grown polypeptide chain. Their presence in a mRNA results in chain termination and signals the end of synthesis of the protein molecule.

Mutations: Changes in the DNA Base Sequence

Once the reading of the mRNA begins at AUG, each triplet is read in order with no interruptions or gaps. If the sequence is changed by even a single base pair (i.e., UUU to UCU), the amino acid specified in the protein will be serine in place of phenylalanine. Thus, a single base change in the DNA sequence leads to a single base change in the mRNA, which may result in a single amino acid change in a protein. If that amino acid change occurs in a part of the protein necessary for its function, the function will be lost. A single base change, or base-pair substitution, in the DNA that results in the replacement of one amino acid in the protein by another is referred to as a missense mutation (Figure **E**). A single base-pair substitution in the DNA also can result in a codon specifying an amino acid being changed to one of the termination codons (i.e., UCG to UAG). When this occurs in the middle of a gene, protein translation stops at that point, and the resulting protein product is a truncated version of the normal product and is thus inactive. A single base-pair substitution in the DNA that leads to the presence of a termination codon in the middle of the coding region of a mRNA is referred to as a nonsense or termination mutation. A third type of mutation can result from an addition or deletion of one or two base pairs in the DNA. When this occurs in the coding region, the reading frame or sequence of triplets read on the mRNA is shifted, resulting in wrong amino acids being placed in the protein from the point of the addition or deletion. Such mutations, in which the mRNA is read out of frame, are known as frameshift mutations. Frameshift mutations usually have drastic effects on a protein, for it will now have multiple amino acids that differ from those found in the normal protein.

Clinical Significance

The process of translation must be extremely accurate. Each tRNA molecule must carry a specific amino acid and contain the proper anticodon sequence to ensure that each protein molecule synthesized will have the exact amino acid sequence that was dictated by the base sequence in DNA. Any deviation, even a single mistake in translation, can lead to a mutant or defective protein. Changes occurring in the DNA base sequence are permanent changes that can lead to missense, nonsense, or frameshift mutations and often result in inherited human diseases.

For more information see Hoffee PA: *IMS: Medical Molecular Genetics.* Fence Creek, pp. 43–47.

12 Protein Synthesis

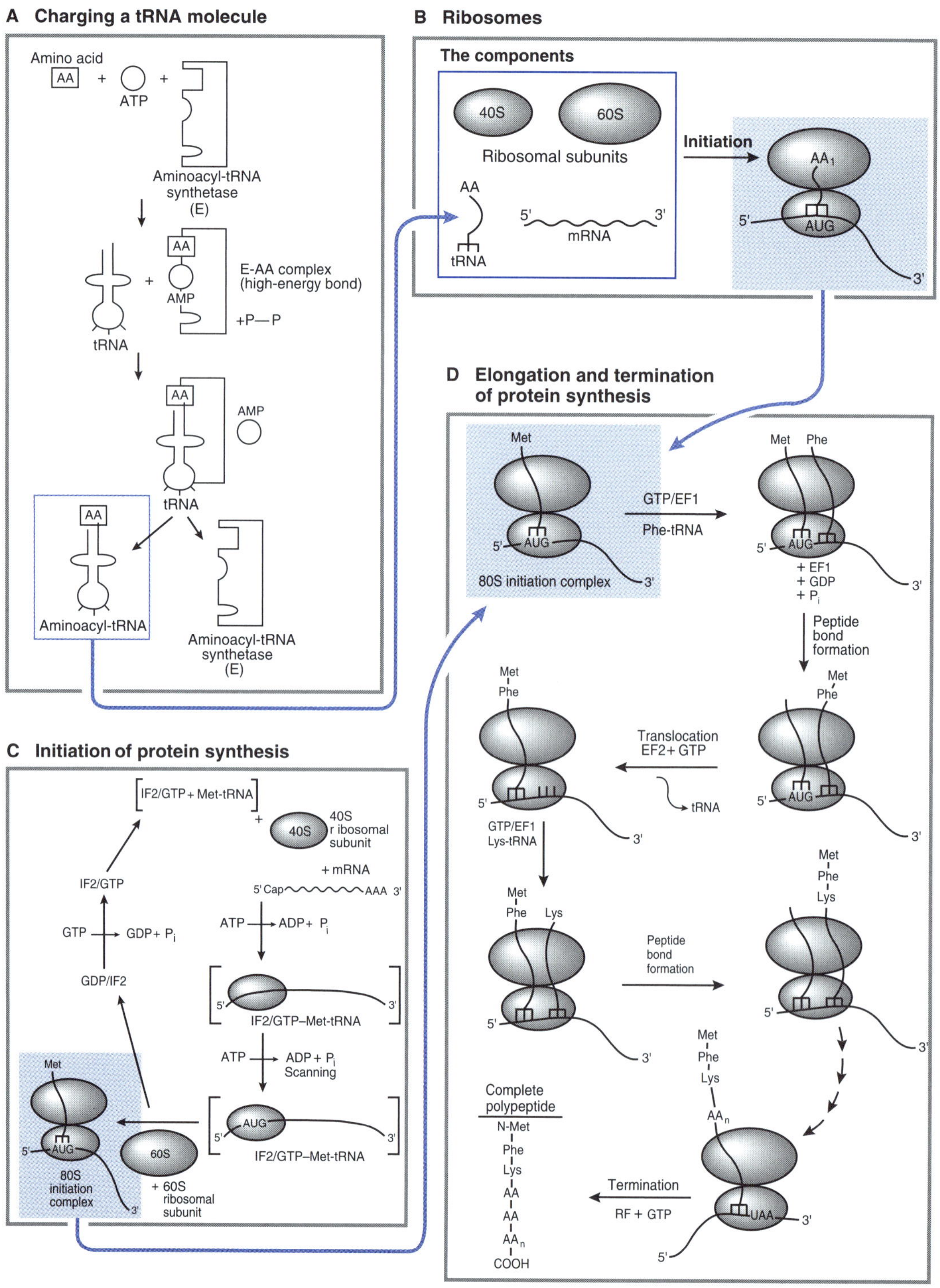

OVERVIEW

The main components of the protein synthesis machinery include mRNA, tRNAs, amino acids, and ribosomes. mRNA provides the sequence of codons that are to be translated into protein. tRNAs bind specific amino acids and then transfer that amino acid into protein, using the specific codons in the mRNA. Finally, a particle consisting of rRNA and protein, known as a ribosome, acts as the site of protein synthesis.

Charging a tRNA Molecule

Charging is the process by which an amino acid is attached to a specific tRNA molecule (Figure **A**). For each of the 20 amino acids found in a cell, there are 20 different enzymes called aminoacyl-tRNA synthetases. The function of these enzymes is to attach a specific amino acid to a specific tRNA molecule, the specificity of the reaction lying in the base sequence of the tRNA molecule. A specific amino acid is initially attached to a specific aminoacyl-tRNA synthetase with the simultaneous hydrolysis of adenosine triphosphate (ATP) to give an enzyme–amino acid–adenosine monophosphate (AMP) intermediate. This intermediate complex binds to a specific tRNA molecule with the corresponding transfer of the amino acid to the 3′ end of the tRNA and the simultaneous release of the enzyme and AMP. The aminoacyl tRNA maintains the energy released in the hydrolysis of ATP in the ester bond formed between the amino acid and the tRNA molecule. This energy will be used later to drive the synthesis of a peptide bond.

Ribosomes

Ribosomes are particles made up of both rRNA and specific proteins. They are found in the cytoplasm, where they serve as the site of protein synthesis. Each ribosome consists of two subunits: the 40S, or small subunit, and the 60S, or large subunit (Figure **B**). When they are joined together, the entire particle is 80S. The 40S subunit is made up of approximately 30 different proteins and contains 18S rRNA. The 60S subunit has 50 different proteins along with the 5S, the 5.8S, and the 28S rRNA molecules. During protein synthesis, the 40S subunit, the 60S subunit, charged tRNA molecules, and a mRNA molecule come together to form a complex that, dictated by the codons in the message, directs the insertion of a specific amino acid into a growing polypeptide chain. A mRNA can be thousands of bases long and is translated by multiple ribosomes, each of which attaches at the 5′ end of the mRNA and moves down the message, producing a single protein molecule. In this way, multiple copies of a single protein can be produced from a single mRNA molecule.

Initiation of Protein Synthesis

Protein synthesis involves three specific stages: initiation, elongation, and termination, each of which requires different proteins and the use of energy in the form either of ATP or guanosine triphosphate (GTP). Initiation utilizes protein factors called IFs, or initiation factors. Multiple IFs are involved in the process of initiation, one of which is IF2. IF2 binds to GTP to form a complex that binds a tRNA charged with methionine (Figure **C**). (Remember that the methionine codon, AUG, is always the first codon or initiator codon on a mRNA molecule.) This complex, IF2-GTP–Met-tRNA, then interacts with the 40S subunit and the 5′ end of a mRNA. After binding to the 5′ end of the mRNA, the 40S subunit, along with the IF2-GTP–Met-tRNA complex, begins to move down or scan the mRNA. This scanning, which uses the 5′ Cap for proper alignment and ATP as an energy source, continues until the complex reaches the first AUG, or initiator codon, on the mRNA. Once the complex finds the first AUG codon, the 60S ribosome subunit binds to the complex to form the final ribosome structure. The attachment of the 60S subunit requires GTP as an energy source. With formation of this final structure, composed of a 40S subunit, a 60S subunit, Met-tRNA, and a mRNA, the initiation step is completed. IF2 and guanosine diphosphate (GDP) are released from the complex, with IF2 being regenerated by GTP to reinitiate the cycle.

Elongation and Termination of Protein Synthesis

Once protein synthesis has been initiated by the binding of the Met-tRNA to the first AUG codon in a mRNA molecule bound to the ribosomes, elongation begins (Figure **D**). The next aminoacyl-tRNA enters the ribosome and binds to the second codon on the message directly following the AUG codon. Before the aminoacyl-tRNA enters the ribosome, it interacts with specific factors known as elongation factors (i.e., EF1), and the energy source GTP. When the tRNA binds to its codon on the mRNA, GTP will be hydrolyzed, with the release of GDP and EF1. EF1 will be regenerated in the presence of GTP and will be able to initiate the cycle once more. On the mRNA, Met-tRNA is bound at AUG; adjacent to it, the second acyl-tRNA, Phe-tRNA, is bound at UUU. In the presence of the enzyme peptidyltransferase, a peptide bond is formed between the carboxyl group of methionine and the amino group of phenylalanine. This reaction results in the attachment of methionine to phenylalanine. In the next step, the ribosome, in the presence of GTP; a second EF, EF2; and an enzyme called translocase, moves (translocates) three nucleotides along the mRNA in a 5′ to 3′ direction. The uncharged tRNA is released, and a new codon on the mRNA is now exposed. This newly exposed codon on the mRNA is in a position to accept the next aminoacyl tRNA, and the cycle is repeated, each step resulting in a growing polypeptide chain with one additional amino acid. When the moving ribosome comes to a termination codon (UUA, UGA, UAG), there is no tRNA that recognizes these codons. Instead, a factor known as release factor, RF, using GTP as an energy source, binds to the ribosome. In the presence of RF, the peptidyltransferase transfers the carboxyl terminal residue of the final amino acid to water, and the peptide is released from the ribosome. The ribosome then dissociates into its 40S and 60S subunits, and these subunits are now ready to begin a new cycle.

Clinical Significance

The process of protein synthesis is extremely complex, yet it must be accurately completed over and over again in a human cell. Any deviation from the normal in even a single step can lead to disease. Protein synthesis is the site of action of numerous antibiotics, an important factor in their use for treating infectious diseases.

For more information see Hoffee PA: *IMS: Medical Molecular Genetics*. Fence Creek, pp. 47–53.

13 Mutations in Human DNA

A Mechanisms by which mutations occur

Mispairing during DNA replication

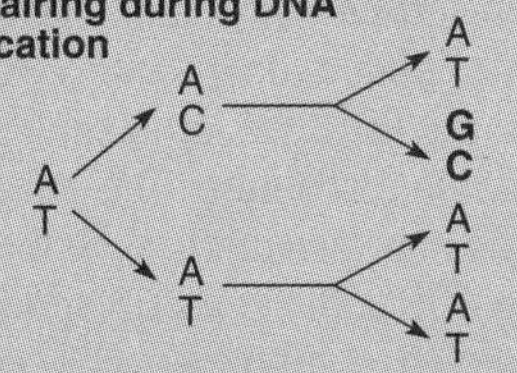

Depurination of adenine or guanine bases

C-A-**G**-T-A ⟶ C-A- -T-A (+**G**)
G-T-C-A-T G-T-C-A-T

Deamination of cytosine, adenine, or guanine bases

C-A-**G**-T ⟶ C-A-**G**-T ⟶ C-A-G-T / G-T-C-A
G-T-**C**-A G-T-**U**-A ⟶ C-A-**A**-T / G-T-**U**-A ⟶ C-A-**A**-T / G-T-**T**-A

Chemical damage

G ⟶ G* (8-hydroxyguanine)
C C

Ultraviolet Irradiation (e.g., sunlight)

C-T-T-G ⟶ C-T^T-G
G-A-A-C G-A-A-C

B Classes of mutations found in human DNA

Normal

DNA	ATGCAGGTGACCTCAGTG TACGTCCACTGGAGTCAC
RNA	AUGCAGGUGACCUCAGUG
Protein	Met-Gln-Val-Thr-Ser-Val

Missense mutation

DNA	ATGCAG**C**TGACCTCAGTG TACGTC**G**ACTGGAGTCAC
RNA	AUGCAG**CUG**ACCUGAGUG
Protein	Met-Gln-**Leu**-Thr-Ser-Val

Nonsense mutation

ATGCAGGTGACCT**G**AGTG
TACGTCCACTGGA**C**TCAC
AUGCAGGUGACC**UGA**GUG
Met-Gln-Val-Thr-**stop**

Single base-pair substitutions (point mutations)

Altered structure of gene product	
Missense mutation	Single amino acid replacement in the protein
Nonsense mutation	A termination codon in the middle of the gene results in premature termination of protein synthesis
RNA-splicing mutation	The protein may be missing part or all of an exon sequence
Altered quantity of gene product	
Mutations in regulatory sequences	Transcription of the gene is altered, which can reduce or eliminate the gene product
Mutations in RNA processing and translation	The stability of the messenger RNA is altered, which may reduce the amount of gene product

Frameshift mutation

DNA	ATGCAGGTG**A**ACCTCAGTG TACGTCCAC**T**TGGAGTCAC
RNA	AUGCAGGUG**AACCUCAGUG**
Protein	Met-Gln-Val-**Asn-Leu-Ser**

Deletion mutation

DNA	ATGTCAGTG TACAGTCAC
RNA	AUGUCAGUG
Protein	Met-Ser-Val

Insertion mutation

ATGCAGGTG-**LINE-3000bp**-ACCTCAGTG
TACGTCCAC-**LINE-3000bp**-TGGAGTCAC
AUGCAGGUG-**LINE-3000bp**-ACCUCAGUG
Met-Gln-Val———?

Insertions or deletions

One or two base pairs (frameshift mutations)	The addition or deletion of one or two base pairs can affect the reading frame of the gene, resulting in a grossly altered or absent gene product
Large numbers of base pairs	Large pieces of the DNA may be lost or large segments of DNA may insert into the middle of a gene, resulting in loss of function

Trinucleotide expansion

DNA	ATG-**(CAGCAGCAG)**$_{20}$CAGGTGACCTCAGTG TAC-**(GTCGTCGTC)**$_{20}$GTCCACTGGAGTCAC
RNA	AUG-**(CAGCAGCAG)**$_{20}$CAGGUGACCUCAGUG
Protein	Met-**(Gln-Gln-Gln)**$_{20}$-Gln-Val-Thr-Ser-Val
DNA	ATG-**(CAGCAGCAG)**$_{75}$CAGGTGACCTCAGTG TAC-**(GTCGTCGTC)**$_{75}$GTCCACTGGAGTCAC
RNA	AUG-**(CAGCAGCAG)**$_{75}$CAGGUGACCUCAGUG
Protein	Met-**(Gln-Gln-Gln)**$_{75}$-Gln-Val-Thr-Ser-Val

Expansion of trinucleotide repeat sequences	Unstable trinucleotide repeats can suddenly expand in number, resulting in the alteration of production or structure of a particular gene product
Chromosomal alterations	Inversions, translocations, duplications, or gene amplification may result

C Clinical significance of mutations in human DNA

Mutation	Disease
Single base-pair substitutions	
Missense mutation	Sickle-cell anemia Glutamic acid (Glu) is changed to valine (Val) in hemoglobin A to produce hemoglobin S
Nonsense mutation	β°-Thalassemia Loss of beta-globin production
RNA splicing mutation	β⁺ -Thalassemia Reduced amount of beta globin
Regulatory mutation	Hemophilia B Reduced transcription of factor IX
RNA translation	α-Thalassemia Reduced amounts of alpha globin
Frameshift mutation	Duchenne muscular dystrophy Altered reading frame from improper splicing of exons gives mutant dystrophin β°-Thalassemia Loss of beta-globin production Tay-Sachs disease Altered reading frame from a 4-bp insertion in exon 11 of the gene coding for hexosaminidase A leads to a deficiency of hexosaminidase A[a]
Small deletion	Cystic fibrosis Deletion of 3 bp at codon 508 in the cystic fibrosis gene (Δ*F508*) leads to a deletion of phenylalanine in the cystic fibrosis transmembrane conduction regulator (CFTR) protein [b]
Large deletion	Duchenne muscular dystrophy No dystrophin protein
Large insertion	Hemophilia A Inactivation of the gene for factor VIII

[a]This mutation is found in 73% of Tay-Sachs disease patients of Ashkenazi Jewish descent.
[b]This mutation, Δ *F508*, is found in 70% of cystic fibrosis patients of Northern European descent.

OVERVIEW

Mutations, defined as permanent changes in the base sequence of a DNA molecule, are the underlying cause of all human genetic disease. All living cells that use DNA as their genetic material have common mechanisms by which mistakes or defects that lead to mutation occur, as well as common systems that prevent these mistakes from remaining in the DNA.

Mechanisms by Which Mutations Occur

During DNA Replication.

1. **Mispairing.** Replication of DNA occurs semiconservatively, with DNA polymerase recognizing a base in the template strand and adding a properly base-paired incoming nucleotide to the 3′ end of the newly synthesized growing DNA strand. Nucleotides, however, are capable of forming an alternative structure, called a tautomer, in which a hydrogen atom is shifted within the molecule, enabling the base to form a mispair (e.g., an A-C pair instead of an A-T pair) (Figure **A**). If the mispair is not corrected by the DNA polymerase, the A-C pair will give rise to a G-C pair and a mutation at the next replication cycle.
2. **Stand-slippage.** When the DNA polymerase encounters a long run of repeated sequences, such as di- or trinucleotide repeats, strand-slippage can occur, resulting in either an increase or a decrease in the number of bases present in the daughter strands. Both mispairing and strand-slippage are repaired by the mismatch repair system.

Alteration of Bases Present in Replicated DNA.

Bases in replicated DNA molecules undergo damage or modification by spontaneous or chemical reactions (see Figure **A**). These reactions include

1. Depurination of an adenine or guanine base, leaving behind a free apurinic site or gap in one of the DNA strands.
2. Deamination of C, A, or G, which results in the loss of an amino group from the base and a change in its structure (i.e., when deaminated C becomes U). U resembles T in its base-pairing properties, and base pairs with an A residue. A GC pair after deamination becomes a GU pair. Following replication, GU becomes AU and eventually becomes an AT pair or a mutation.
3. Chemical modification of compounds in the cell, which leads to damage by producing intermediates that oxidize bases in DNA, causing interference with normal replication or transcription. Spontaneous and chemical changes to bases in DNA are repaired by the base excision repair systems.
4. Radiation damage, which occurs when agents present in the environment, such as ultraviolet (UV) rays from sunlight, produce bonding between adjacent pyrimidines on the same DNA strand. The most frequent type of dimer produced is the thymine–thymine dimer, which, when present in a DNA molecule, blocks DNA replication and leads to the death of the cell. Radiation damage is repaired by the nucleotide excision repair system.

Classes of Mutations Found in Human DNA

Mutations in DNA molecules affect the structure of the product of a gene or the amount of the product coded for by a gene (Figure **B**).

1. Many of the mutations that occur in DNA are caused by a single base-pair substitution in which one base pair, such as an A-T pair, is replaced by a second base pair, such as a G-C pair. The substitution of one base pair by a second base pair results in the change of a codon and can lead either to a missense mutation, in which one amino acid replaces another amino acid in a protein, or to a nonsense mutation, in which one of the termination codons appears in the middle of a gene. When a nonsense codon (UAA, UGA, or UAG) appears in the middle of a gene, protein synthesis terminates at the site of the nonsense codon on the mRNA, and a truncated polypeptide is produced. Single base-pair substitutions also occur in sequences other than those coding for amino acids. For example, a mutation that alters the splice acceptor or splice donor sequences can result in aberrant splicing of the RNA, giving rise to a mRNA missing part or all of an exon sequence. Other base-pair substitutions occur in regulatory sequences that are necessary for the binding of TFs or RNA polymerase. These mutations can result in a reduced quantity of the product produced by the gene, a complete absence of the gene product, or an increase in the amount of the gene product.
2. Frameshift mutations are caused by the addition or deletion of one or two base pairs within the coding sequence of a gene, leading to alteration of the reading frame of the mRNA. The sequence of codons in a mRNA is translated by the protein-synthesizing machinery, starting at the first AUG codon and then consecutively reading the mRNA in groups of three nucleotides without interruption. By addition or deletion of one or two base pairs in the DNA, a mRNA is produced with an altered sequence and is now translated out of frame from the site of the insertion or deletion of the base pair. Frameshift mutations result in the production of a protein that will be altered in its amino acid sequence, starting from the point of the insertion or deletion of the base pair and continuing to the end of the protein.
3. Insertion and deletion of many base pairs also can occur within DNA molecules. Deletion mutations may occur in a chromosome, with the loss of hundreds to thousands of base pairs from the DNA, in which case the deleted genetic material is permanently lost from the cell. Insertion mutations (large insertions of DNA sequences), are rare occurrences in human cells and are caused by transposon-like elements, often repetitive DNA sequences such as the long interdispersed nuclear element (LINE) repeat.
4. Mutations due to strand-slippage result in the expansion of trinucleotide repeats (CAG, CTG, CGG, or GAA). Runs of these repeat triplets occur in exons at the 5′ end or the 3′ end of genes and in some instances even within an intron. Normal healthy individuals have a particular copy number range and, in most cases, have different numbers of repeats in their maternal and paternal alleles. Individuals affected with one of the expansion disorder diseases have a sudden increase in the number of copies of the trinucleotide repeats, giving rise to an alteration in either the expression or the structure of a particular protein or RNA molecule (see Chapter 30).

Clinical Significance

Examples of some human inherited diseases that arise from the different classes of mutations in DNA are listed in Figure **C**.

For more information see Hoffee PA: *IMS: Medical Molecular Genetics*. Fence Creek, pp. 58–64.

14 DNA Repair I

Repairing Mistakes Made During DNA Replication

A Proofreading by DNA polymerase

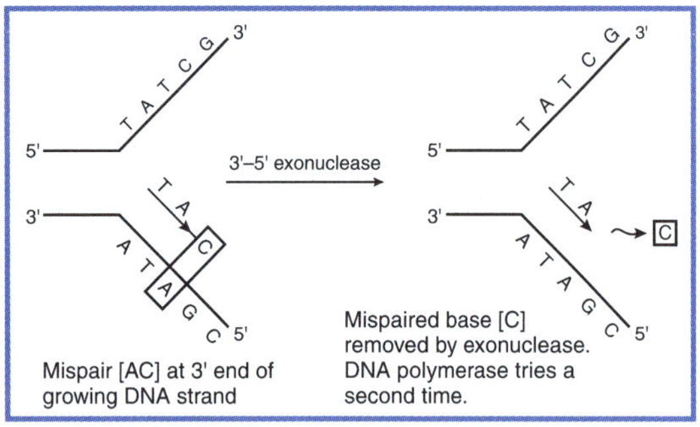

B Methyl-directed mismatch repair (MMR) system

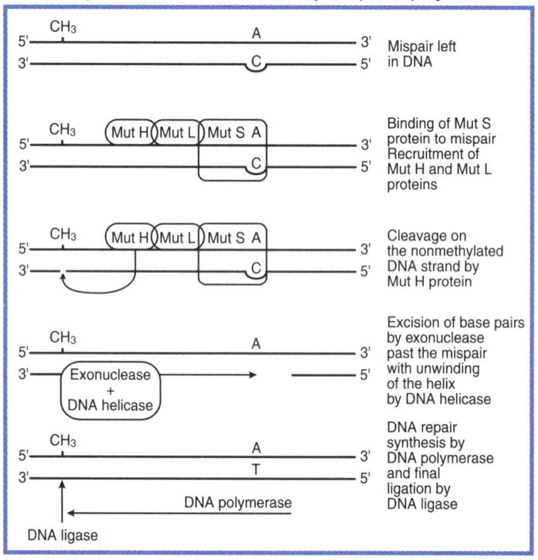

C HNPCC: A result of a defective MMR system

Proteins involved in MMR		
Human Homologues	***Escherichia coli***	***Function (E. coli)***
hMSH2	Mut S	Damage recognition
hMSH3	Mut S	
hMSH6	Mut S	
hMLH1	Mut L	Complex formation
hPMS1	Mut L	
hPMS2	Mut L	
?	Mut H	Endonuclease

D Trinucleotide expansion disorders

		Repeats		
Disease	***Protein***	***Normal***	***Mutant***	***Type***
Huntington's disease[a]	Huntingtin	11–34	36–120	Glutamine (CAG) repeats
DRPLA[a]	Atrophin	7–23	49–75	
Spinobulbar muscular atrophy[a]	Androgen receptor	11–33	40–62	
Spinocerebellar ataxia (type I)[a]	Ataxin-1	6–44	40–82	
Spinocerebellar ataxia (type 3)[a]	MJD-1	13–40	68–69	
Fragile X syndrome[b]	FMR-1	6–50	60–1000	CGG repeats
Myotonic dystrophy[c]	Protein kinase	5–30	50–2000	CTG repeats
Friedrich's ataxia[d]	Frataxin	10–21	200–900	GAA repeats

[a] Glutamine repeats (CAG) in the coding region of the first exon of the gene.
[b] CGG repeats in the noncoding 5' end of the first exon of the gene.
[c] CTG repeats in the noncoding 3' end of the last exon of the gene.
[d] GAA repeats within the first intron of the gene.
DRPLA = dentatorubral-pallidoluysian atrophy; FMR-1 = originally for fragile X mental retardation; MJD = protein defective in spinocerebellar ataxia type 3, which is also known as Machado-Joseph disease.

OVERVIEW

DNA polymerase replicates 6.6 billion bp of human DNA once in every S phase. Although the process of DNA replication is quite accurate, mispairs occur at a frequency of 1 in 1 million bp replicated. Yet, the mutation rate in a cell is not 1 in 1 million bp but is closer to 1 in 1000–10,000 million bp replicated. To keep the mutation rate low, cells have two mechanisms that deal with the occurrence of DNA mispairs made by DNA polymerase during DNA replication: proofreading by DNA polymerase, and the methyl-directed mismatch repair (MMR) system.

Proofreading by DNA Polymerase

Associated with DNA polymerases that carry out DNA replication is an enzyme activity known as a 3′ to 5′ exonuclease. Human DNA polymerase-δ and -ε both have this activity. The 3′ to 5′ exonuclease activity is responsible for cleaving a mispaired nucleotide from the 3′ end of the newly replicated strand of DNA (Figure **A**). When an incoming nucleotide in the altered tautomeric form mispairs with a base in the template strand, that mispair may result in a distortion in the DNA. The 3′ to 5′ exonuclease sees this as a mistake and cleaves the nucleotide from the 3′ terminus. The DNA polymerase then has a second chance to put the proper nucleotide in the new DNA strand. With the accuracy of the DNA polymerase and the associated 3′ to 5′ exonuclease activity, the fidelity of DNA replication reduces the mutation rate to 1 in 10^9 bp replicated.

Methyl-Directed Mismatch Repair System

Mispairs that occur during DNA replication are not always recognized by the 3′ to 5′ exonuclease, and the replication fork moves past the mispair, leaving it in the newly replicated DNA. If the mispair remains in the DNA, it leads to a mutation at the next replication cycle. To correct these mispairs, cells have evolved a repair system, the MMR system, that removes mispairs immediately after replication of the DNA. The MMR system was initially discovered during the study of *Escherichia coli* cells that exhibited a mutator (*mut*) phenotype, in which cells show an increased rate of spontaneous mutation at all genetic loci. A similar repair system has now been described for human cells. The MMR system constantly scans the DNA, looking for mispairs as well as small insertions and deletions that are not properly paired to the template DNA strand and give rise to single-stranded loops of DNA. When a mispair is detected, the repair system removes the defect from the DNA, and the correct base sequence is restored (Figure **B**). For proper correction to occur, the repair system must be able to recognize which DNA strand needs to be repaired. For example, if the normal base pair is A-T, and the mispair is A-C, correction and removal of the A from the mispair will lead to a G-C pair, or a mutation. On the other hand, removal of the C from the mispair will give back the original A-T, or correct base pair. Recognition is dependent on the absence of methylated bases on the newly replicated DNA strand. In human cells, cytosine residues located in CpG islands are methylated; however, methylation is a postreplication event, and newly synthesized DNA is not immediately methylated. Therefore, during the initial period of DNA replication, one strand—the template strand—is methylated, but the newly synthesized DNA strand is not methylated. When a mispair is detected, correction specifically takes place on the nonmethylated, or newly synthesized, DNA strand. The proteins known to be involved in MMR are referred to as mutator (Mut) proteins. In *E. coli*, these proteins are Mut S, Mut L, and Mut H. Human homologues are known for Mut S (hMSH2, hMSH3, and hMSH6) and Mut L (hMLHI, hPMS1, and hPMS2), but at this time none are known for Mut H (Figure **C**). In the process of DNA repair by the MMR system, the mismatch is first recognized as defective by the Mut S protein (see Figure **B**). The presence of the Mut S protein bound to the mismatch recruits the Mut L and Mut H homologues to the site. The next step is a cleavage of the nonmethylated DNA strand opposite the methylated site by Mut H, followed by the stepwise removal of nucleotides by an exonuclease as far as some 100 bp past the mismatch. This step is facilitated by a DNA helicase that unwinds the DNA duplex. The resulting gap in the DNA is then repaired by DNA polymerase, using the base sequence in the template strand. The final step is the sealing of the final phosphodiester bond by DNA ligase. Additionally, nucleotides that are not base paired and that loop out from the DNA are also repaired in human cells by the MMR system.

Hereditary Nonpolyposis Colorectal Cancer (HNPCC)

One important finding relating defective DNA repair to cancer is the discovery that a common hereditary cancer, HNPCC, is caused by mutations in genes that code for proteins involved in the MMR system. HNPCC affects 1 in 200 persons in the United States and accounts for 15% of all colon cancers. Evidence accumulated over the past few years indicates that at least six genetic loci are involved in the human MMR process (Figure **C**). Proteins coded for by these genes are homologues of the *E. coli* proteins and function in a similar way to repair mispairs in human DNA. The hallmark of HNPCC is microsatellite instability, which presumably is a result of the inability to remove displaced loops from the DNA. Microsatellite instability leads to a *mut* phenotype—that is, an increased rate of overall mutation and an increased chance of developing mutations in tumor suppressor genes.

Trinucleotide Expansion Disorders

The disorders listed in Figure **D** result from the expansion of trinucleotide repeat sequences (CAG, CTG, CGG, or GAA) that are found in exons at the 5′ or 3′ ends of genes. The number of these repeats is unstable and can vary from one individual to another. Individuals affected with one of the disorders have a sudden increase in the number of copies of the trinucleotide repeats, which gives rise to an altered RNA or protein molecule. Expansion of the repeats can occur during replication when strand-slippage by DNA polymerase produces single-strand loops of DNA that are not properly repaired by the MMR system.

For more information see Hoffee PA: *IMS: Medical Molecular Genetics*. Fence Creek, pp. 64–67.

15 DNA Repair II

Repairing Damaged or Altered Bases in DNA

A Base-excision repair system

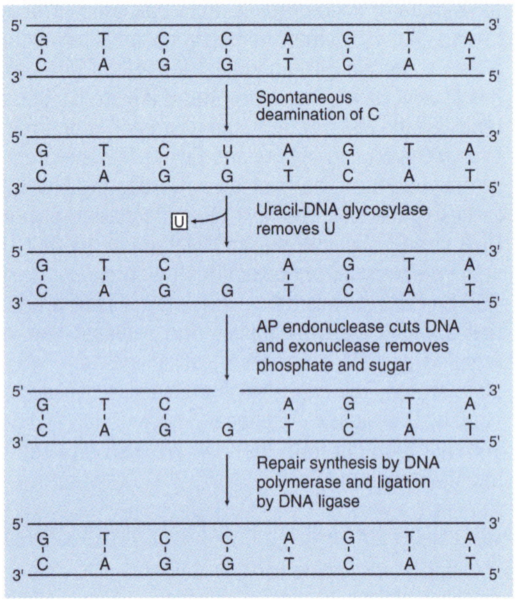

B Nucleotide-excision repair system (NER)

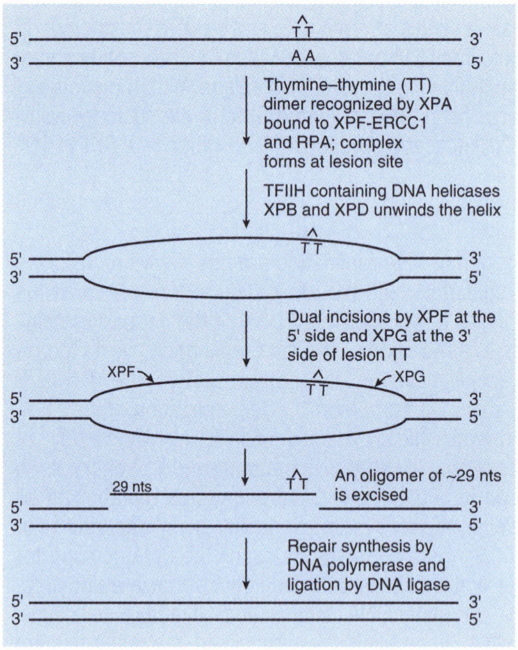

C Defects in nucleotide excision repair resulting in human disease

Genes	Function
Xeroderma pigmentosum (XP) genes	
XPA	Damage recognition
XPB/XPD	Helicase activity
XPC	ATPase/DNA binding
XPE	Damage recognition
XPF	5' SS endonuclease
XPG	3' SS endonuclease
Cockayne syndrome (CS) genes	
CSA	Unknown
CSB	TFIIH Subunit
Tricothiodystropy (TDD) genes	
TTDA	TFIIH Subunit
XPB/XPD	Helicase activity

OVERVIEW

In addition to errors occurring during DNA replication, another source of mutation is the alteration or damage of bases in a DNA molecule that is not undergoing DNA replication. Cells have evolved two major repair systems to deal with this type of DNA damage or alteration: the base-excision repair system and the nucleotide-excision repair (NER) system.

Base-Excision Repair System

Uracil, a normal component of RNA, is similar to thymine in its base-pairing properties with adenine. Uracil, however, is rarely found in DNA. This specificity seems even more puzzling, since the precursor for putting uracil into DNA (i.e., dUTP) is a good substrate for DNA polymerases, and at least in vitro, it can be used to synthesize DNA that contains uracil. The key to this puzzle came when a new enzyme called uracil-DNA-glycosylase was described in bacteria. This enzyme removes uracil residues from DNA by cleaving the glycosidic bond between the base and the deoxyribose residue. Uracil-DNA-glycosylase is also present in human cells and is responsible for keeping uracil out of DNA molecules. Other enzymes have now been described that have a similar glycosylase activity; that is, they act in cells to remove specific altered or damaged bases from the DNA by the pathway diagramed in Figure **A**. Uracil residues arise in DNA primarily by the spontaneous deamination of cytosine residues. When a uracil residue appears in a DNA molecule, it is recognized by uracil-DNA-glycosylase, and it is removed from the DNA, leaving behind a gap. The lack of a base in the DNA, which is called an apurinic or apyrimidinic site (AP site), is recognized by specific endonucleases known as AP endonucleases, which cleave the DNA strand at the site of the missing base. The resulting gap is extended by an exonuclease and then repaired by DNA polymerase, using the base present in the complementary strand as a template. The final step is a ligation by DNA ligase to give back a corrected DNA strand. A cell must remove the uracil residues from DNA to prevent the G-U mispairs formed by cytosine deamination from being converted to A-T pairs or a mutation at the next replication cycle. An additional problem arises when cytosines in the DNA are methylated, as they often are. Deamination of methylated cytosine gives rise to a thymine in the DNA. The presence of thymine, although it will be mispaired to G, is not recognized as a wrong base in DNA. (Remember that this is happening in DNA, which is already replicated and methylated, so the methyl-directed mismatch repair system will not correct this G-T mispair.) As a result, sites in the DNA where cytosine is methylated are prone to high mutation rates and are referred to as hot spots for mutation.

Nucleotide-Excision Repair System

A more general repair system that is especially important for repairing many kinds of damage that occur to DNA molecules is known as the NER system. This repair system is necessary for the repair of bulky distortions in the DNA molecule. The NER system is the most complex system known for DNA repair and involves a large number of proteins. The NER process, diagramed in Figure **B**, shows the steps that occur during the repair of DNA that has been damaged after exposure to UV light or sunlight. UV light damages DNA by causing dimers to form between adjacent pyrimidines on the same DNA strand, the most common dimer being a thymine–thymine (T-T) dimer. When a T-T dimer is present in the DNA, distortion of the helix causes replication and transcription to stop at that point until the dimer is removed by the NER system. The initial step is recognition of the dimer by the protein XPA, which binds to the lesion along with the XPF-ERCC1 protein complex and the RPA or single-stranded DNA-binding protein. A complex of TFIIH, which contains the DNA helicases XPD and XPB as well as other polypeptides, is recruited to the site by the presence of XPA. The helicase activity unwinds the helix and stimulates the excision activity of two endonucleases, XPF and XPG. The two nucleases make a duel excision, with XPG cutting the DNA 3 nucleotides 3′ to the damage and XPF cutting the DNA 24 nucleotides 5′ to the damage. The result of dual excision is the release of an oligomer of 29 nucleotides containing the T-T dimer, leaving behind a DNA molecule with a large gap and a free 3′ hydroxyl end. The 3′ hydroxyl is recognized by DNA polymerase-δ or -ε, which now carries out repair synthesis, using the undamaged DNA strand as a template. The final nick is sealed by DNA ligase, and the DNA is now repaired.

Defects in NER Resulting in Human Disease

Figure **C** indicates the genes in human cells that code for the proteins involved in NER. These genes and their functions have been identified by the study of patients with different mutations, all of which lead to the autosomal recessive disease Xeroderma pigmentosum (XP) or to two related autosomal recessive diseases: Cockayne syndrome (CS) and trichothiodystrophy (TTD). The genes are named as *XPA, XPB*, etc., indicating a defect in a gene (i.e., gene *A* or gene *B*) that results in the disease XP, or *CSA, CSB*, indicating a defect in a gene *A* or *B* that leads to CS. Patients with XP are very susceptible to skin cancers upon exposure to sunlight because they lack one or more of the proteins needed to carry out the process of NER. Patients with CS have growth and mental retardation and also are sensitive to sunlight, but they do not develop skin cancer at an increased rate. These patients generally are of short stature and have progressive neurologic degeneration. Patients with TTD generally have brittle hair, short stature, mental retardation, and some neurologic problems. As with CS patients, they do not develop skin cancer at an increased rate. Mutations in the *XPB, XPD*, and *XPG* genes also lead to TTD and may indicate a defect in the coupling of repair to transcription.

For more information see Hoffee PA: *IMS: Medical Molecular Genetics*. Fence Creek, pp. 66–71.

16 DNA Damage Control by p53

A Damage control by p53

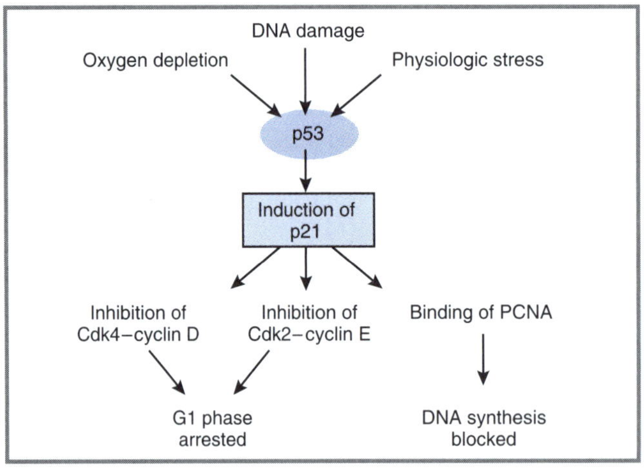

B DNA damage results in a block in G1

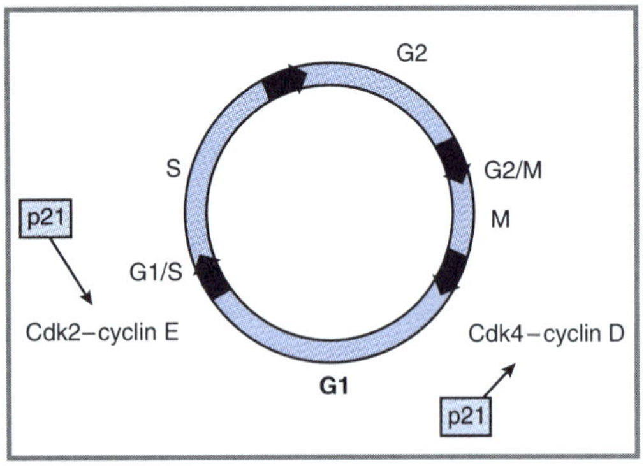

C p53 potential therapy for cancer

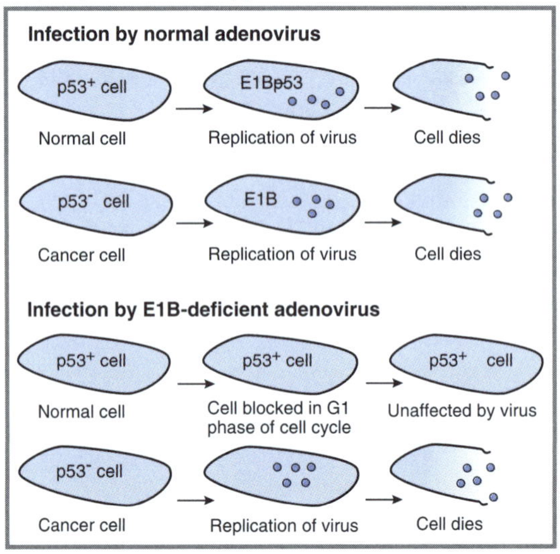

OVERVIEW

The extensive systems of DNA repair in cells emphasize the importance of protecting the integrity of the genome. Otherwise, damage to the DNA or changes in the base sequence would accumulate, leading to mutations, possibly to cancer, and sometimes to the death of the cell. In addition to these repair pathways, cells have other safeguards to ensure that DNA damage is corrected before the DNA is replicated and transmitted to daughter cells.

Damage Control by p53

Crucial to normal growth and development is the ability of a cell to sense some deviation from the normal. Problems arising in a cell, such as damage to the DNA, a lack of oxygen, a lack of sufficient energy or metabolites, infection by viruses, or any physiologic disturbance that can affect the normal growth and cycling of the human cell must be dealt with or an abnormal cell can result. To counteract such problems, cells have evolved negative regulatory mechanisms that can sense these abnormal conditions and either arrest a cell at a particular stage of the cell cycle or in some cases direct the cell to apoptosis (Figure **A**). The p53 protein, which is a DNA-binding protein that acts as a transcription factor (TF), is the key controlling element that sits at the center of these negative regulatory mechanisms. p53 is a tumor suppressor protein, and any mutations that result in the loss or alteration of its normal activity can lead to a high rate of cancer development. Cells that do not have a functioning p53 lose the ability to stop the progression of the cell cycle when adverse conditions arise. The cell continues to proliferate, usually leading to a defective cell type. p53 exerts one of its controls by activating transcription of the gene coding for the Cdk inhibitor, p21, resulting in increased levels of the p21 protein.

DNA Damage Results in a Block in G1

The p21 protein functions by binding to multiple cyclin–Cdk complexes and blocking their kinase activity, thus blocking the phosphorylation of proteins needed for progression in the cell cycle (Figure **B**). The binding of p21 to the G1 cyclin–Cdk complexes, Cdk4–cyclin D and Cdk2–cyclin E, is central to the G1 arrest that follows DNA damage by ionizing radiation. An arrest of the cell in G1 allows time for the DNA repair systems to correct the damage. In some cases, the damage can be repaired by the systems described in Chapters 14 and 15; in other cases, the damage is too extensive, and the cell will be directed to apoptosis. Another function of p21 is to bind proliferating cell nuclear antigen (PCNA), which is a factor required to activate DNA polymerase-δ. With p21 bound to PCNA, DNA replication is inhibited, thus preventing the damage from being replicated before it is repaired. Thus, the ability of a cell to stop progression in the cell cycle to allow repair of damaged DNA to occur is an important control process that prevents the occurrence of mutations that might lead to a cancerous cell.

p53, a Potential Therapy for Cancer

More than 50% of all human cancer cells are defective in p53. This finding has led to a new approach to treat cancer by directly killing the tumor cells that are defective in p53. The procedure relies on the use of defective adenoviruses that cannot replicate in cells expressing the tumor suppressor protein p53 (Figure **C**). Normal adenoviruses contain a gene that codes for a protein called E1B. During the infection process, the E1B protein binds to the p53 protein and stimulates the cells to enter the S phase of the cell cycle and support DNA replication of the virus. Thus, viruses have evolved proteins like E1B to block the function of p53 and stimulate cells to bypass the G1 block and enter the S phase of the cell cycle. In theory, an adenovirus mutant that lacks the E1B protein should be unable to replicate in cells that express the p53 protein. In the presence of a functioning p53, the virally infected cells are blocked in the G1 phase of the cell cycle and will not enter into the S phase; thus, the virus is unable to replicate its DNA, and the infection is aborted. However, a cell lacking the p53 protein should serve as a receptive host for the defective adenovirus because the cell is no longer blocked in G1 in response to the viral DNA; thus, it progresses to the S phase and allows the defective virus to replicate and kill the host cells. Because over 50% of human cancer cells are defective in p53, a mutant adenovirus that does not express E1B could potentially be used to target these cancer cells and specifically kill them. Any normal cells present should use their p53 protein to prevent the replication of the defective virus and should remain viable. Recently, this theoretical approach to selectively kill cancer cells defective in p53 has been demonstrated to be feasible. A mutant adenovirus defective in the E1B protein was shown to lyse p53-deficient human cells but not kill cells that expressed a normal p53 protein. Furthermore, when the mutant virus was injected into human cervical carcinomas present in nude mice, there was significant regression of the tumor size and complete regression in 60% of the injected tumors. Currently, clinical trials are under way to test this therapy in individuals with head or neck cancers who have not responded to other therapies. Whether or not this therapy will work in humans is yet to be determined. The major obstacle in humans is the presence of an immune system, which potentially can prevent the spread of the virus in the tumor. The therapy currently is limited to tumors that can receive direct injections, but even if it works on a few cancers it will be an important and beneficial treatment.

Clinical Significance

Tumor suppressor proteins are crucial in a cell to prevent the development of cancers. p53 is a tumor suppressor protein that plays a critical role in cell cycle control, particularly in response to DNA damage. Allowing the cell to pause in the G1 phase of the cell cycle in order for the DNA repair processes to carry out correction of the damaged DNA appears to be a critical process in the prevention of cancer. This damage control role of p53 in preventing cancer is supported by the finding that over 50% of all human cancers are defective in p53.

For more information see Hoffee PA: *IMS: Medical Molecular Genetics*. Fence Creek, pp. 346–347.

17 Regulation of Gene Expression I

Cis-Acting Elements and *Trans*-Acting Transcription Factors

A ***Cis*-acting elements in a human gene**

Cis-acting sequence	*Transcription factor*
Core promoter elements	
Inr	RNA polymerase II*
TATA	TATA-binding protein (TBP)*
Promoter proximal elements	
Common elements	
CAAT	CTF
GC	Sp 1
Octamer	OTF
Response elements (RE)	
GRE	Glucocorticoid receptor
TRE	AP1
CRE	CREB
Promoter distal elements	
Enhancers	Activators
Silencers	Repressors

* GTFs bind to RNA polymerase and TBP at the core promoter

B ***Cis*-acting elements in a human gene**

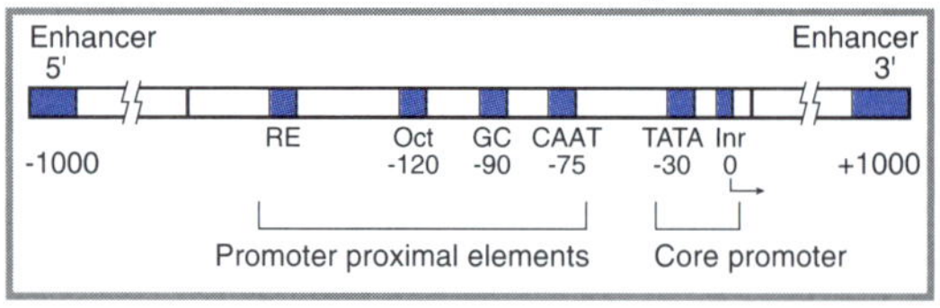

C ***Trans*-acting transcription factors**

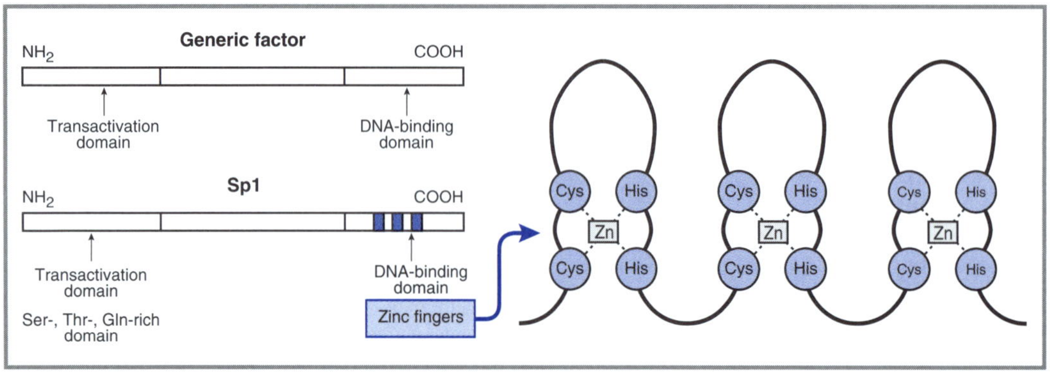

D **Classes of *trans*-acting transcription factors**

Class of TFs	*Function*
General	Form preinitiation complex
Common	Bind common sequences (CAAT)
Inducible	Bind RE sequences
Activators	Bind enhancers
Repressors	Bind silencers

OVERVIEW

Every cell in a human body contains the same DNA sequence, a sequence that carries information for 60,000–100,000 different RNA or protein molecules. Yet, not every cell in the body expresses the same genes. Even within a single cell, different genes are expressed at different times in response to external stimuli, developmental controls, or cell cycle signals. Cells must have the capacity to regulate the turning on and turning off certain genes at the right time. For example, some proteins are required only at a certain phase of the cell cycle, and their expression at other times would be detrimental. Any deviation from the normal control of the cell cycle leads either to uncontrolled cell proliferation and malignancy or to programmed death. Additionally, cells must be able to regulate the expression of specific genes in response to hormones or growth factors to carry out certain functions required in the life and development of an organism. The predominant step in all cells for the regulation of gene expression is at the level of transcription initiation, the process in which RNA polymerase binds to a specific site on the DNA and begins to transcribe an RNA molecule from a DNA template. Transcriptional control, which occurs mainly by controlling the access of RNA polymerase to the promoter site of a gene, is an amazingly complex process that involves the interaction of proteins (i.e., *trans*-acting factors) with specific sequences on the DNA (i.e., *cis*-acting elements).

Cis-Acting Elements

Cis-acting elements are short base-pair sequences that can be adjacent to or within a gene, or they can occur thousands of base pairs away from the gene. These sequences are required for the recognition of a gene by RNA polymerase II, and they act as binding sites for the proteins that regulate the rate and the specificity of transcription. The major types of *cis*-acting elements of a gene are listed in Figure **A**, and their positions within a gene are diagramed in Figure **B**. The core promoter elements are the initiator element (Inr), which determines the transcription start site, and the TATA element, located 25–30 bp upstream of the Inr. These sequences are located 5′ to the gene and constitute the site where the transcription initiation complex containing RNA polymerase II assembles on the DNA. The promoter proximal elements are found 50 to a few hundred base pairs upstream of the start site and include two types: (1) A common class of base sequences (e.g., CAAT or GC) that occur in many genes and act as binding sites for proteins called upstream transcription factors (TFs). (2) The response element (RE) that occurs in promoters under common control (e.g., the genes that respond to glucocorticoid stimulation). The promoter distal elements are *cis*-acting sequences that can occur thousands of base pairs away from the transcription start site and are known as enhancers or silencers. These may be located either 3′ or 5′ to the gene they regulate.

Trans-Acting Transcription Factors

Most proteins that function as TFs have two major features in common: they bind to specific *cis*-acting elements on the DNA, and they interact with other TFs to regulate transcription initiation. Analysis of the structure of TFs indicates that they generally are composed of two functional domains (Figure **C**). One part of the protein, the DNA-binding domain, is the portion of the molecule that recognizes and binds specific DNA sequences. The ability of proteins to bind DNA is a reflection of particular amino acid sequences in the protein and the formation of specific motifs, one of which is the zinc finger motif. The zinc finger motif is found in several TFs, including the common TF, Sp1, the general transcription factor (GTF), TFIIA, and the glucocorticoid receptors. This motif is named for the characteristic structures that are formed when zinc binds to cysteine and histidine residues to form finger-like protrusions that can interact with DNA. The second important functional domain of TFs, known as the transactivation domain, is the amino acid sequence in the protein that is necessary for the activation of transcription. The transactivation domain, which is a serine-, threonine-, glutamine-rich amino acid sequence, is the part of the protein that interacts with other protein factors such as the TFIIs or RNA polymerase II, the proteins that make up the basal transcription complex.

Classes of *Trans*-Acting Transcription Factors

Figure **D** lists main classes of TFs, including the following: GTFs are a set of 30 or more polypeptides that assemble at the core promoter site and recruit RNA polymerase II to that site to form a preinitiation complex. One of these polypeptides is the TATA-binding protein, TBP, which is a subunit of TFIID. Common TFs, sometimes called upstream TFs, are proteins (e.g., Sp1) that bind the common *cis*-acting sequences proximal to many promoters, such as the sequences CAAT and GC. Inducible *TFs* are proteins that are not always active in the cell. They respond to external stimuli that activate them and potentiate their binding to the RE sequences. Only when the inducible TF is activated in the cell is transcription of genes containing a particular RE element activated. Activator proteins are TFs that bind enhancers and increase transcription initiation of a gene. Repressor proteins are TFs that bind silencers and inhibit transcription initiation of a gene.

Clinical Significance

Crucial to normal growth and development of all cells, as well as to the organism as a whole, is the proper programmed expression of every gene in every cell. All cells need to express the proteins necessary for providing energy and synthesizing the components necessary for growth. By contrast, some proteins are expressed only in certain cells. Examples include the synthesis of globin, which is generally confined to the erythroid cells, or the synthesis of immunoglobulins, which occurs predominantly in B lymphocytes. A normal cell, and thus a normal individual, is dependent on the proper gene regulation occurring in every cell. Deviation from this normal regulation leads to many human disease conditions such as the hemoglobinopathies, immune dysfunction, and hormonal dysfunction, to name just a few.

For more information see Hoffee PA: *IMS: Medical Molecular Genetics*. Fence Creek, pp. 164–167.

18 Regulation of Gene Expression II

Transcription Initiation

A GFTs and their functions

GFT	*Number of subunits*	*Putative function/properties*
TFIIA	3	Stabilizes TFIID-DNA complex; responds to activators
TFIIB	1	Promotes TFIIF-RNA polymerase II interaction; start site selection
TFIID	10	TBP binds TATA sequence; TAFs interact with activators and repressors; TAFIID 250 is a serine kinase
TFIIE	2	Stimulates TFIIH CTD kinase activity
TFIIF	2	Protein kinase; escorts RNA polymerase II to the preinitiation complex
TFIIH	9	Helicase activity; cyclin-dependent kinase activity; cyclin; CTD kinase activity

Note: CTD = carboxyl-terminal domain

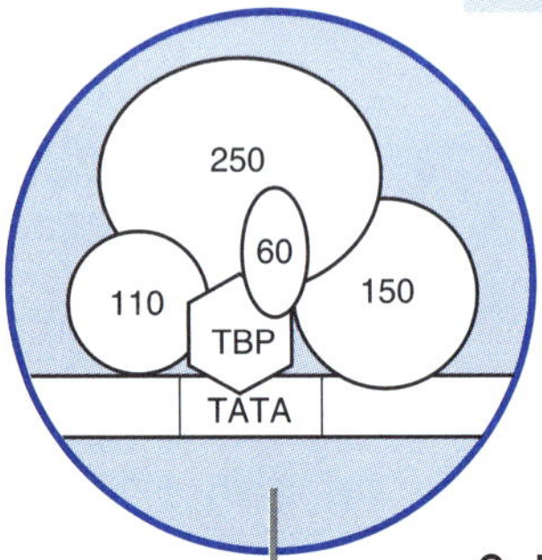

B TFIID

TATA-binding protein (TBP) and TBP-associated factors (TAFs)

TAFIID 250, TAFIID 150, TAFIID 110, TAFIID 60

C Initiation of transcription by RNA polymerase II

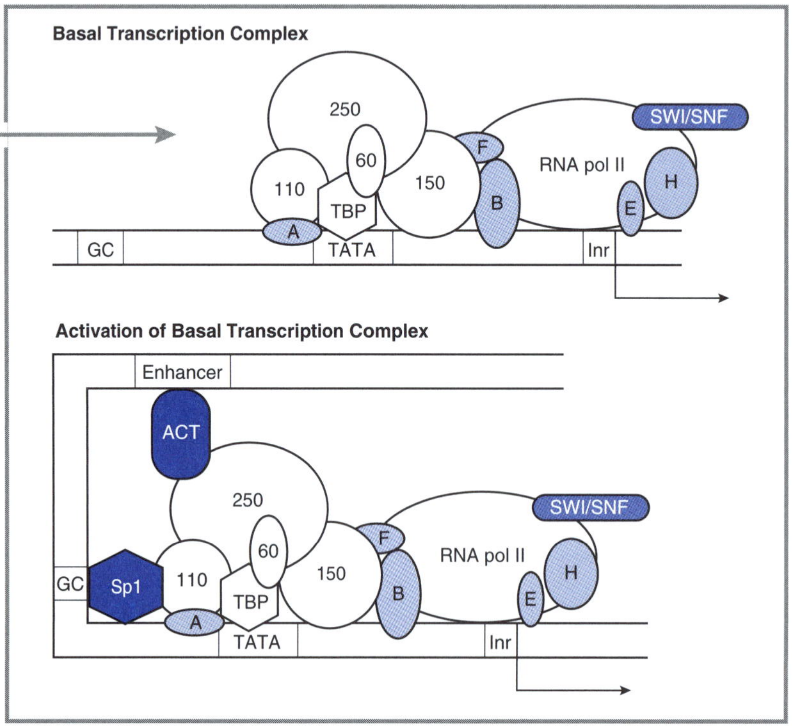

OVERVIEW

Transcription initiation by RNA polymerase II is a complex process that requires the precise interaction of multiple proteins for proper regulation of the expression of each gene. The GTFs are the proteins that assemble at the core promoter and recruit RNA polymerase II to the Inr site to form the preinitiation complex. This complex is further activated and regulated by the other types of TFs, such as the common TFs, the inducible TFs, and activator proteins.

General Transcription Factors and Their Functions

GTFs, listed in Figure **A**, are named TF for transcription factor and II for RNA polymerase II control, followed by a letter (A, B, D, E, F, or H) to indicate different factors. Many of the GTFs known to be involved in the initiation of transcription have been purified and characterized, and their putative function in transcription initiation has been studied using in vitro transcription systems. TFIID is the first factor to bind to the core promoter sequence (Figure **B**). TFIID is made up of at least 10 polypeptides, one of which is the TATA-binding protein (TBP). TBP, a protein of 38 kD, specifically recognizes the TATA sequence at all promoter sites. When TBP binds to the DNA, the protein forms a saddle-like shape around the DNA double helix. This distorts and bends the DNA and is believed to facilitate the binding of additional proteins. The remaining polypeptides in TFIID are referred to as TBP-associated factors (TAFs). The TAFs range in molecular weight from the largest of 250 kD (TAFIID 250) to the smallest of 10 kD (TAFIID 10), and they interact with TBP to form a complex at the TATA-binding site. Functions attributed to the various TFII proteins (in addition to their ability to interact with TF and control the initiation of transcription) include helicase activity, protein kinase activities, and cyclin-like activity.

Initiation of Transcription by RNA Polymerase II

Transcription initiation requires the interaction of RNA polymerase II and multiple TFs, forming a basal transcription complex at the core promoter site of a gene (Figure **C**). Once TFIID is bound at the promoter by way of the binding of TBP to the TATA sequence, the other GTFs bind to the complex in the following order—TFIIA, TFIIB, TFIIF/RNA polymerase II, TFIIE, and TFIIH—to form the preinitiation complex. When this complex is stimulated by other TFs, mRNA synthesis begins with the movement of RNA polymerase II away from the promoter, and elongation of the mRNA transcript. The efficient and specific recognition of the core promoter site by the basal transcription complex depends on short *cis*-acting sequences that are located upstream of the TATA sequence. These sequences include the common sequences found in RNA polymerase II promoters such as CAAT, Oct, and GC. Specific upstream TFs recognize these sequences and bind to the DNA. As indicated in the lower panel of Figure **C**, the TF Sp1, which contains a zinc finger DNA-binding domain, binds to GC sequences, known as GC boxes. Once bound to the DNA, Sp1 can interact directly with the basal transcription complex through protein–protein interactions by way of its transactivating domain. The transactivating domain of Sp1 recognizes the TAFIID 110 subunit of TFIID bound at the core promoter, and, by some still unknown mechanism, enhances the assembly of the GTFs and RNA polymerase II at the promoter site. Other upstream-binding factors, such as AP1 and OTF, presumably work in a similar manner; that is, once bound to the DNA, they can activate transcription, although each protein may interact with a different factor in the basal transcription complex. The regulation of transcription initiation by enhancer and silencer sequences poses an interesting question of how sequences located thousands of base pairs away from a gene can control the transcription of that gene and do it in such a way that the sequence is independent of orientation (i.e., it can be upstream or downstream of the initiation start site). To regulate transcription initiation, the TFs bound at enhancer sequences must be brought into proximity to the basal transcription complex. The most likely hypothesis to explain the mechanism by which a distant sequence can interact at a promoter site is the process of DNA looping, in which the DNA located between the promoter site and the enhancer sequence loops out to bring the two sequences close together, as indicated in the lower panel of Figure **C**. In this way, the proteins bound at the enhancer sequences and the TFs bound at the promoter site can interact directly to activate transcription initiation. Silencers are thought to act by a similar mechanism, except that the binding of a repressor protein to a silencer would inhibit the transcription initiation process.

Role of Chromatin Structure in Transcription Regulation

Inside the cell, the DNA is found wrapped around histones in the form of nucleosomes and ordered chromatin structure (see Chapter 3). For transcription initiation to occur within a natural chromatin structure, the protein factors necessary to carry out the process must gain access to the promoter sites on the DNA. If a promoter site is accessible to proteins, then transcription can occur, but if a promoter site is inaccessible and blocked by histones, the gene remains in a nontranscribed state. Two findings have provided a glimpse of how the chromatin structure might be disrupted to expose promoter sites and allow transcription to occur. One is the acetylation of lysine residues at the amino terminal end of histones, resulting in a destabilization of the nucleosome structure and exposure of the DNA to proteins involved in transcription initiation. The TBP-associated protein, TAFIID 250, contains a histone acetyltransferase activity and may be important in altering the chromatin structure. In addition, RNA polymerase II contains a multiprotein complex known as the SWI/SNF complex (Figure **C**). The SWI/SNF complex, which is made up of at least 10 proteins, includes a protein with ATPase activity that, in the presence of ATP, disrupts the nucleosome array. This disruption can facilitate the transcription of DNA that was previously unavailable to the transcription complex.

For more information see Hoffee PA: *IMS: Medical Molecular Genetics*. Fence Creek, pp. 167–170.

19 Regulation of Gene Expression III

Inducible Control of Transcription

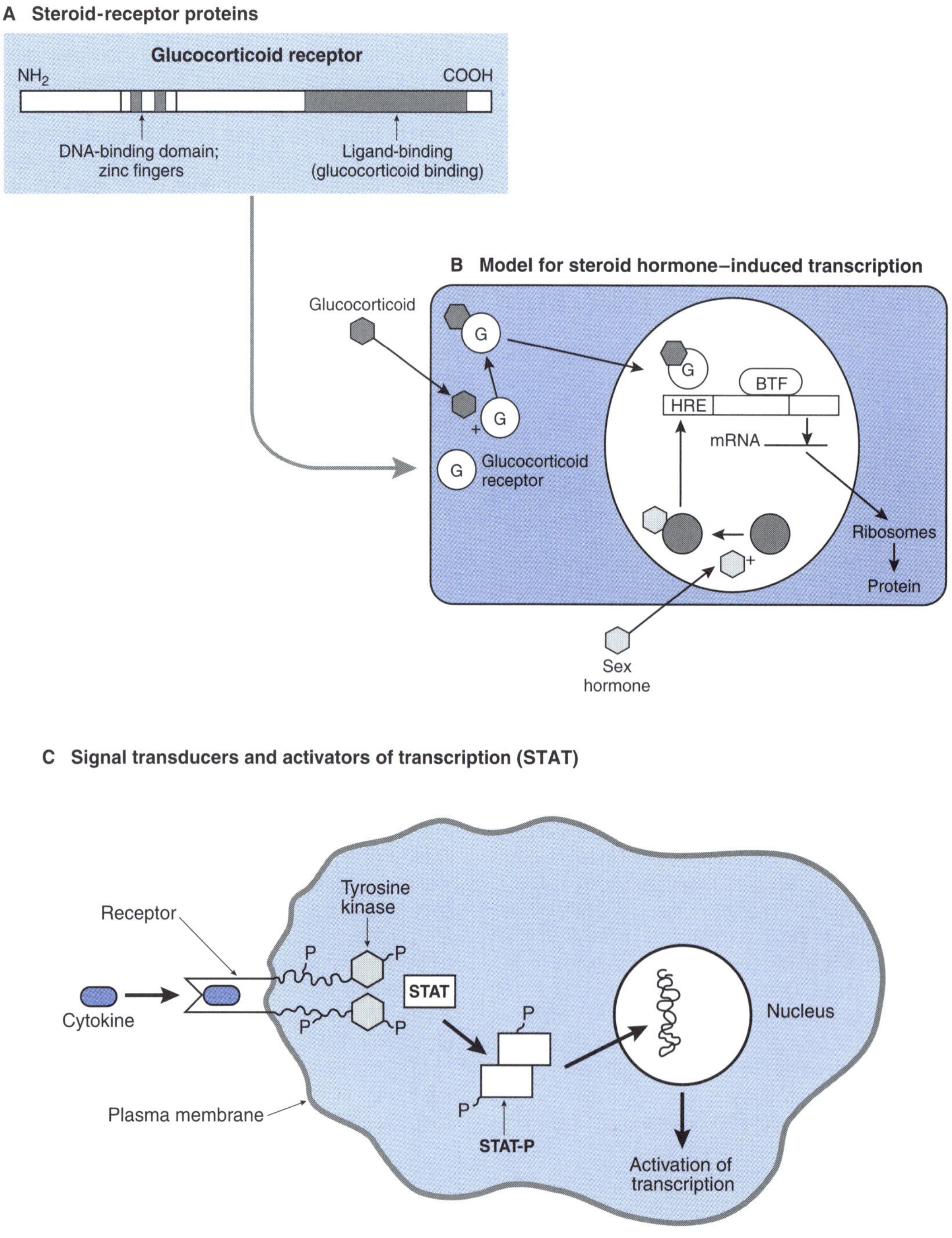

OVERVIEW

Some classes of genes are under common control in the sense that their transcription is activated in a similar manner under specific conditions. For example, the presence of glucocorticoids or cytokines in the external milieu of a cell results in a specific induction or increased transcription of all of the genes induced by these compounds. This inducible response is attributed to the *RE sequences*. These upstream *cis*-acting sequences occur only in special promoters and act as binding sites for specific inducible TFs.

Steroid Receptor Proteins and Hormonal Control of Transcription

An important example of inducible transcription is the response of cells to steroid hormones. Proper growth, development, and homeostasis in humans is controlled in many instances by steroid and steroid-like hormones. Steroid hormones, such as glucocorticoid and mineralocorticoid, are important for regulating glycogen and mineral metabolism. The sex steroids (estrogen, progesterone, and testosterone) are essential for normal sexual differentiation and reproduction. Other important biologic processes require steroid-like elements. For example, vitamin D is necessary for bone development, and retinoic acid functions as a regulator of differentiation. Steroid and steroid-like hormones carry out their functions by binding to specific proteins, the steroid receptors, to form an activated complex that is now capable of binding to RE sequences found in specific genes. The activated steroid-receptor proteins are TFs that, when bound to RE sites in the DNA, activate transcription of a specific class of genes. Steroid-receptor proteins have several common features, which include a DNA-binding domain and a hormone-binding domain (Figure **A**). As seen in many other TFs, the DNA-binding domain of steroid receptors contains zinc finger motifs.

Model for Steroid Hormone–Induced Transcription

Steroids are synthesized in the endocrine cells and move to their target cells through the blood. Because of their lipid solubility, they can enter cells by diffusion across the plasma membrane (Figure **B**). Once inside the cell, the steroid binds to a steroid receptor present in either the cytoplasm or the nucleus. Some steroid receptors (e.g., the glucocorticoid receptor) are present in the cytoplasm, whereas others (e.g., the androgen receptor) are found in the nucleus. A glucocorticoid enters a cell and is bound by the glucocorticoid receptor protein present inside the cytoplasm of the cell; then, the complex moves to the nucleus. Sex hormones, however, must enter the nucleus before they can bind to the androgen receptor. In both cases, the binding of the hormone activates the receptor protein so it is now capable of binding to DNA. Once in the nucleus, the activated complex binds to the RE sequences on the DNA known as the hormone response elements (HREs). The HRE sequences may be located upstream of the promoter or sometimes in enhancer sequences. The binding of the activated complex to the HRE, and the subsequent interaction of the receptor protein with factors in the basal transcription complex (BTF), result in activation of the initiation of transcription by RNA polymerase II. All genes that have a common HRE sequence will be simultaneously activated by a specific activated complex and show increased transcription. In this way, a cell can coordinate the inducible expression of multiple genes in response to specific hormone signals.

Signal Transducers and Activators of Transcription (STATs)

When a water-soluble ligand, such as a growth factor or a cytokine, is involved in activating transcription, the receptor protein for the ligand is localized within the plasma membrane. One important class of membrane protein receptors is made up of those that have tyrosine kinase activity and react to growth factors or cytokines to regulate cell growth. Also of importance in these pathways are STATs, which are TFs present in an inactive form in the cytoplasm (Figure **C**). When cytokines bind to receptors present in the plasma membrane, a tyrosine kinase associated with the receptor phosphorylates the receptor, providing a binding site for STAT proteins. The bound STAT proteins undergo phosphorylation on tyrosine residues and form dimers, which then migrate into the nucleus. Once inside the nucleus, they act as TFs by binding to specific sequences on the DNA located upstream of the TATA sequence. Although each cytokine or growth factor has a different set of genes that it activates, they all act through a limited number of STATs. Currently, six STAT family members are known. How this differential expression of genes occurs is still unknown and will require a knowledge of how the different STATs interact with each other and with other TFs bound at specific *cis*-acting sequences.

Clinical Significance

The precise regulation of genes under inducible control is essential to normal growth and development. This regulation involves many steps, including the production of the ligands, the binding of the ligand and the activation of the inducible TF, and the stimulation of transcription initiation by the binding of the TF to the proper RE site on the DNA. Any deviation in even one of these steps often leads to cancer, as discussed in Chapter 21. Additionally, mutations that occur in the X-linked gene coding for the androgen receptor result in a male with androgen insensitivity, a syndrome also known as testicular feminization. Affected individuals occur with an incidence of 1 in 20,000 and have a normal male karyotype (46,XY). However, they have female genitalia and female characteristics but have a blind vagina and no uterus. Secretion of androgen by the testes, which are located in the abdomen, is normal in these individuals. However, in the absence of the androgen receptor, transcriptional regulation of the genes required for normal male fetal development is lost, and the characteristics associated with normal maleness do not develop.

For more information see Hoffee PA: *IMS: Medical Molecular Genetics*. Fence Creek, pp. 170–172

20 Regulation of Gene Expression IV

Developmental Control of the Globin Genes

A Alpha-globin and beta-globin gene clusters

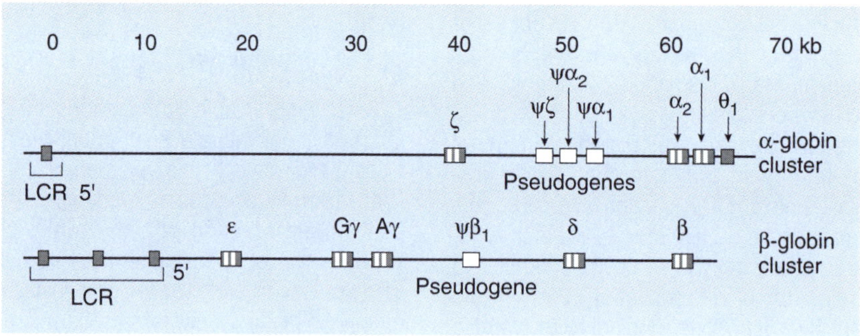

B Hemoglobin subunit composition during development

Hemoglobin subunit composition	*Designation*	*Developmental period*
z_2e_2	Hb Gower 1	Embryonic
z_2g_2	Hb Portland	Embryonic
a_2e_2	Hb Gower 2	Embryonic
a_2g_2	HbF	Fetal
a_2b_2	HbA	Adult, major form
a_2d_2	HbA_2	Adult, minor form

Note: Hb = hemoglobin; *HbA* = adult hemoglobin; *HbF* = fetal hemoglobin

C Globin synthesis at various stages of development

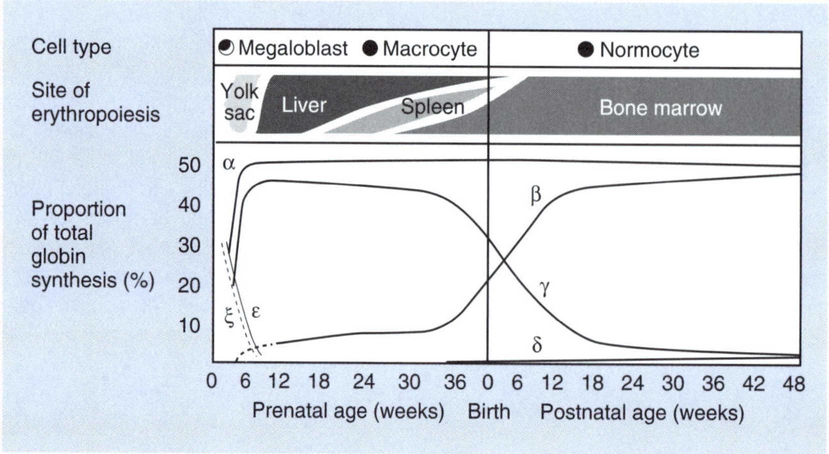

OVERVIEW

The understanding of the regulation of hemoglobin (Hb) synthesis has been fundamentally important to human biology and human genetics. Studies on hemoglobinopathies, the most common single-gene disorders in humans, have contributed a wealth of knowledge to the molecular basis of human disease. In addition, the unique regulated pattern of synthesis of the globin genes during human development has features of special interest.

Alpha-Globin and Beta-Globin Gene Clusters

Normal Hb, a tetramer made up of two alpha-like and two beta-like subunits, functions to carry oxygen in red blood cells (RBCs). The genes that code for the alpha-like subunits and the beta-like subunits of Hb are found in two gene clusters located on separate human chromosomes (Figure **A**). The genes coding for the beta-like–globin chains are found on the short arm of chromosome 11 (11p15.5) and span approximately 50,000 bp of DNA. The beta-like–globin genes code for the epsilon, the G gamma, the A gamma, the delta, and the beta subunits of Hb. The genes coding for the alpha-like–globin chains are located on the short arm of chromosome 16 (16p13.11–13.33) and cover approximately 30,000 bp of DNA. The alpha-like–globin genes code for the zeta and alpha chains. The two alpha genes, α_1 and α_2, are expressed equally and code for identical polypeptides. In addition, this cluster includes three pseudogenes (genes that do not code for a protein) and a theta gene with an unknown function.

Hemoglobin Subunit Composition during Development

Figure **B** gives the subunit composition of the different Hb synthesized during development. In the adult, the major Hb present is HbA. HbA_2 also is found in adults and makes up about 3% of the total Hb present. Fetal cells have HbF, whereas embryonic cells contain three different hemoglobin molecules, Hb Gower 1, Hb Gower 2, and Hb Portland. The presence of different forms of Hb during development requires that different genes in the two clusters be turned on or off at precise times. This process, known as hemoglobin switching, allows the synthesis of the proper Hb subunits at the proper state of development. Additionally, the site of synthesis of Hb during development differs.

Globin Synthesis at Various Stages of Development

Early during development, Hb Gower 1, Hb Gower 2, and Hb Portland are synthesized in the yolk sac (Figure **C**). At about 8 weeks of gestation, HbF and a small amount of HbA begin to be synthesized in the liver. By 18 weeks of gestation, synthesis of Hb begins in the bone marrow with a switch in gene expression that progressively results in a decrease in synthesis of the gamma chain and an increase in synthesis of the beta chain. By the end of the first year of age, HbF is present as less than 2% of the total Hb, with the main Hb present as HbA. The mechanisms responsible for determining which globin genes are expressed during the embryonic, the fetal, and the adult stages of human development is determined by regulatory sequences within the promoter of each globin gene and a common regulatory sequence known as the locus control region (LCR) (see Figure **A**). The LCR, located 6–40 kb upstream of the structural genes, is required both for tissue-specific control of the globin genes and for determination of which genes within the locus will be expressed at a particular time during development. Within the beta-globin gene locus, a temporal expression of the genes occurs from the 5′ to 3′ direction with respect to the position of the LCR. This directionality results first in the expression of the epsilon gene, an embryonic gene, in development, then in the expression of the gamma genes, or the fetal genes, and finally in the expression of the beta gene, or the adult gene. The order of the globin genes with respect to the position of the LCR appears to be an important aspect of control for their temporal expression. The LCR for the beta-globin gene locus, located 6 kb upstream of the epsilon gene, has multiple sites, or *cis*-acting elements, necessary for the binding of TFs. Two of these TFs are GATA-1 and NFE-2, factors that are expressed only in erythroid cells providing tissue-specific regulation of the beta-globin genes. Additionally, AP1 and Sp1, common TFs, also bind to sequences in the LCR. The binding of TFs at the LCR may alter the chromatin structure around the beta gene locus, allowing additional TFs access to the promoters of the individual genes in the cluster. The switch in tissue expression is most likely dependent on the presence of specific TFs in each tissue. For example, embryonic RBCs in the yolk sac have TFs that permit the LCR to activate the transcription of the epsilon gene. Switching of Hb synthesis from the yolk sac to liver cells is accompanied by switching the synthesis of the embryonic form of Hb to the fetal form of Hb. This switch in transcribing the gamma gene and not the epsilon gene depends on the presence in hepatocytic cells of a repressor that binds to a silencer and represses transcription of the epsilon gene. In addition, progenitor fetal RBCs have specific TFs that recognize the promoter of the gamma genes and activate their transcription. A LCR is also present 40 kb upstream of the zeta gene in the alpha-globin cluster. This LCR presumably functions like the one present in the beta-globin gene cluster to control the expression of the zeta and alpha genes.

Clinical Significance

Just about every possible type of mutation in the globin genes and its relationship to human disease have been described. Examples of these disorders of Hb include a missense mutation in the beta-globin gene, giving rise to sickle cell anemia; nonsense, deletion, or frameshift mutations that result in the absence of one of the globin chains (i.e., β^{0}-thalassemia); and RNA-splicing mutations or mutations in regulatory sequences that result in reduced levels of the globin chains (i.e., β^{+}-thalassemia).

For more information see Hoffee PA: *IMS: Medical Molecular Genetics.* Fence Creek, pp. 172–174.

21 Abnormal Transcriptional Control

Oncogenes and Cancer

A Stimulation of growth and cell division in a normal cell

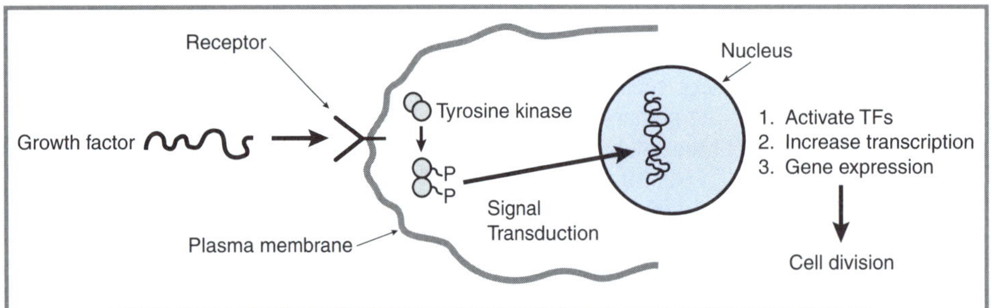

POINTS OF MUTATION THAT CONVERT PROTO-ONCOGENES TO ONCOGENES

1. Increase the amount of growth factor
2. Mutate receptor so no signal is required to activate
3. Mutate the regulatory proteins (tyrosine kinase, *ras*)
4. Mutate transcription factors to change function

B Examples of oncogenes and oncoproteins

General class	*Oncogene*	*Oncoprotein*
Protein kinases		
Membrane-associated proteins	*src, abl*	Tyrosine protein kinase
Transmembrane proteins	*erbB, neu, fms*	Tyrosine protein kinase
Cytoplasmic protein	*mos*	Serine/threonine protein kinase
Growth factors	*sis*	B-chain platelet-derived growth factor
	int-2	Fibroblast growth factor
Signal transducers (GTP-binding proteins)	*H-ras, K-ras, N-ras*	GTPases
Nuclear factors	*jun, fos*	Activator protein-1
	myc, myb	DNA-binding proteins
	bcl-1	Cyclin-D

C Mechanisms that activate proto-oncogenes

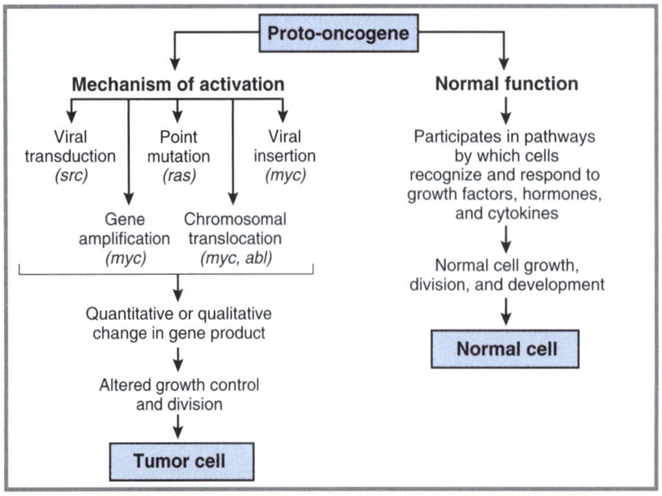

OVERVIEW

During the past 20 years, studies on the molecular basis of cancer have shown that interference with the functions of normal pathways that cells use to recognize and respond to growth factors, hormones, cytokines, and other regulatory molecules can result in the development of a cancer cell. Two important types of genes, oncogenes and tumor suppressor genes, play key roles in determining whether a cell will remain under normal controls or whether it will develop aberrant control signals.

Oncogenes

An oncogene is a viral or cellular gene that can induce neoplastic transformation or cancer when introduced into a normal cell. Oncogenes arise at the single somatic cell level and act in a dominant fashion, but they are not inherited through the germ line. All cellular and viral oncogenes originate by alteration of a normal cellular gene, known as a proto-oncogene, and thus alteration of the function of a normal gene product in the cell. The understanding of oncogenes and their role in causing cancer has led to increased information about many key pathways that a cell uses in normal growth and development. A simple scheme depicting the stimulation of growth of a normal cell by growth factors is shown in Figure **A**. Growth factors bind to receptors on the plasma membrane. This binding can activate the receptor, often a tyrosine kinase, resulting in phosphorylation and activation of the tyrosine kinase or other regulatory molecules and stimulation of the signal transduction pathways. The activated molecule moves to the nucleus, where it modifies TFs, resulting in increased transcription and expression of the genes required for cell division. Alterations in the genes coding for the proteins involved in this general pathway can lead to altered proteins or altered levels of proteins. These altered genes are called oncogenes, and the altered proteins are called oncoproteins.

Examples of Oncogenes and Oncoproteins

Oncogenes code for oncoproteins that have protein kinase activity, that act as growth factors, that function in the signal transduction pathways, or that function as nuclear factors (i.e., TFs) (Figure **B**). Tyrosine protein kinases phosphorylate proteins on tyrosine residues and play a key role in signal transduction pathways. These proteins often are found either spanning the plasma membrane or associated with the plasma membrane of cells. Tyrosine kinases can act as growth factor receptors and are important to the cell, enabling it to respond to growth factors and to relay the growth factor signal to the transcriptional machinery. The binding of a growth factor to a receptor stimulates the tyrosine kinase activity, often resulting in autophosphorylation and activation of the receptor. Many of the known oncogenes code for mutant forms of a normal tyrosine kinase, in which the mutation alters the regulation of the kinase protein in such a way that it no longer requires a ligand (e.g., a growth factor) for activation. In these cases, the receptor is always on, in the sense that it signals a message independently of the presence of the controlling ligand. Growth factors are the key ligands for cell cycle control, and they signal a cell to enter the G1 phase of the cell cycle from a resting phase or to bypass the restriction point in G1. Important for the normal growth of a cell is the proper amount of a growth factor produced at the right time, either in the growth cycle or in the differentiation of a cell. Two oncogenes that code for growth factors include the *sis* oncogene, which codes for the B chain of the platelet-derived growth factor (PDGF), and the *int-2* oncogene, which codes for a fibroblast-related growth factor. Tumor cells express and respond to these oncoproteins, resulting in noncontrolled growth, whereas their normal cell counterparts do not express the growth factor and thus remain quiescent until the growth factor is provided by the appropriate cell at the appropriate time.

An example of an oncogene that affects signal transduction is *ras*. Studies on *ras* have shown that it codes for a GTP–binding protein with GTPase activity. Normal GTP-binding proteins, or G proteins, are important in signal transduction pathways, in which the protein exists in an equilibrium state between an active protein with GTP bound and an inactive protein with GDP bound. In response to the appropriate signal, the inactive form of the G protein is converted to the active form by the exchange of GDP with GTP. The normal proto-oncogene is converted to the *ras* oncogene by a single base-pair substitution that results in the production of a missense mutant protein, which remains in the GTP bound state and thus is always activated. Many oncogene products act in the nucleus as aberrant TFs. Oncogenes in this class include the *jun* and *fos* oncogenes. In a normal cell, the proto-oncogenes *fos* and *jun* code for subunits that make up the inducible TF, AP1. Included in this class of nuclear oncogenes are those that code for proteins that function in the progression of the cell cycle, such as the oncogene *bcl-1*, which is a result of a mutation in the normal gene coding for cyclin D.

Mechanisms That Activate Proto-Oncogenes

Activation of proto-oncogenes to oncogenes generally involves two main types of alterations: a quantitative change leading to an increase in the amount of a normal gene product present in a cell, or a qualitative change that involves a genetic mutation within a gene, resulting in the production of an abnormal gene product (Figure **C**). A proto-oncogene can be activated to become an oncogene by several different mechanisms: viral transduction, point mutations within a host gene, viral DNA insertion into the host DNA, gene amplification of host DNA sequences, or chromosomal translocations. In all cases, the quantitative or qualitative change in a particular gene product results in altered growth control and cell division and leads to a cancerous cell.

For more information see Hoffee PA: *IMS: Medical Molecular Genetics*. Fence Creek, pp. 174–177.

22 Tumor Suppressor Genes

A Inheritance pattern of a defective tumor suppressor gene

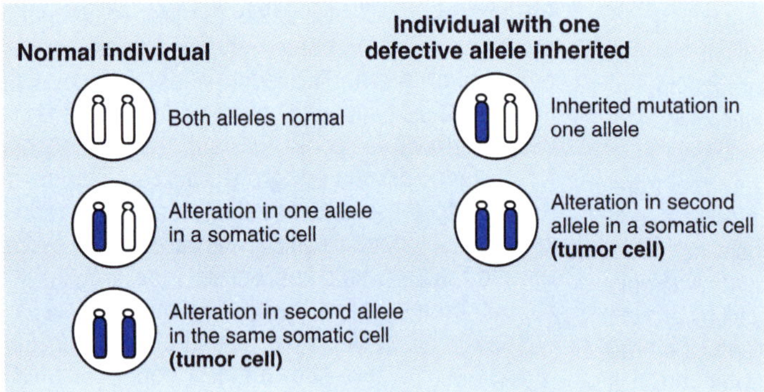

B Two-hit hypothesis

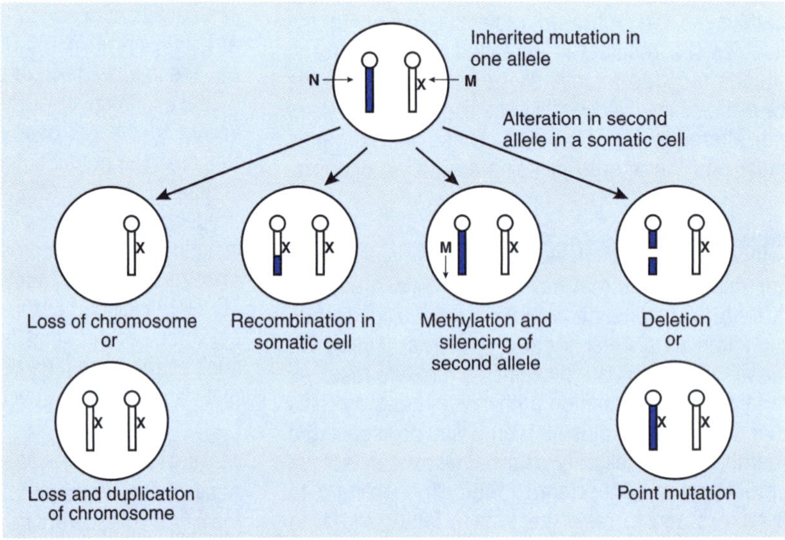

C Examples of human tumor suppressor genes

Chromosome location	*Gene/Protein*	*Tumor/Disease*
3p26	VHL (subunit of elongin)	Von Hippel-Landau disease
5q21	APC	Adenomatous polyposis coli
9p21	Cyclin-dependent kinase inhibitor, p16	Familial melanoma
11p13–15	TF WT1	Wilms' tumor
11p15.1	TSG101	Breast cancer
13q14	Rb protein	Retinoblastoma
13q12	BRCA2	Early-onset breast cancer
17p13	p53	Li-Fraumeni syndrome
17q21	BRCA1	Early-onset breast cancer
17q11.2	Neurofibromin (GTPase)	Neurofibromatosis 1
22q12	Schwannomin	Neurofibromatosis 2

OVERVIEW

The regulation of normal cell growth and division is subject to the action of positive activators, which, in response to the proper stimuli, activate cytoplasmic products or the transcription of specific genes. Mutations in the genes that code for the positive activators often lead to the production of an altered gene product, an oncoprotein. Another important aspect of the regulation of normal cell growth is the role of gene products that function by negatively controlling cell proliferation. The normal growth of cells is a balance between regulation by positive activators, which stimulate growth, and regulation by negative-acting factors, which inhibit growth. In either case, a mutation in the genes coding for these factors can lead to the loss of growth regulation and the development of cancer. When a mutation occurs in a gene that acts in a negative fashion and the product of the gene is either lost or altered, negative regulation is also lost, and uncontrollable cell growth occurs. Genes whose loss or inactivation leads to uncontrollable cell proliferation are known as tumor suppressor genes. The products of tumor suppressor genes act in several different ways to suppress tumor formation. They are important for regulating cell cycle progression, functioning to stop or start the growth of a cell at a particular phase in the cell cycle. They can act by regulating the transcription of specific genes, either by activating transcription of some genes or by repressing transcription of other genes. Tumor suppressor proteins also can direct a cell to undergo differentiation or to undergo a programmed death.

Inheritance Pattern of a Defective Tumor Suppressor Gene

Mutations in tumor suppressor genes are, for the most part, loss-of-function mutations, in which the mutation in the gene results in the absence or loss of a functional gene product. These mutations are recessive; therefore, mutations in both alleles of the gene must occur before a mutant phenotype is observed (Figure **A**). The recessive nature of tumor suppressor mutations allows the heterozygous state to be inherited through the germ line because a single defective allele does not interfere with the development of the fetus. When one mutant allele of a tumor suppressor gene is inherited, the affected individual does not inherit cancer but inherits a predisposition to develop cancer. If a mutation in the second allele of the tumor suppressor gene occurs, it occurs at the somatic cell level, and that single cell will become cancerous. In a normal individual, both alleles of a tumor suppressor gene are active and code for a functioning protein product. Therefore, two separate mutations must occur in a single cell to inactivate both alleles and produce a cancerous cell. In an individual who has inherited one defective allele, every cell in the body carries that defect, and only one mutation at the somatic cell level is necessary to produce a totally defective or cancerous cell. Therefore, an individual who inherits one defective allele of a tumor suppressor gene has a greater chance of developing cancer than a normal individual and is said to have inherited a predisposition to develop cancer.

Two-Hit Hypothesis

The identification and crucial role that tumor suppressor genes play in the regulation of cell growth came from studies on retinoblastoma (RB), an inherited disorder expressed in children as tumors of the retina. Studies on the inheritance patterns of RB by Knudson led to the "two-hit hypothesis" to explain the inherited nature of RB (Figure **B**). According to this hypothesis, a single mutation occurs in the germ line and inactivates one allele of a crucial gene. This mutation is passed on to the offspring of an affected parent. A second mutation occurs in a somatic cell of the offspring, knocking out the other allele to give a homozygous state and a predisposition of the cell to develop cancer. This hypothesis proved correct for RB when it was shown that a single gene at 13q14 was inactivated in families that have the inherited form of RB. This loss-of-function mutation would be expected to be inherited as an autosomal recessive trait, and in fact this initial mutation does not in itself lead to the development of cancer and is recessive to the normal allele. However, cells with this single allele knocked out have a greater probability of developing cancer if another mutation knocks out the second allele to give a homozygous state. There are several ways in which the second allele can be inactivated. This process, termed loss of heterozygosity (LOH), can occur by (1) loss of the chromosome with the normal allele or the loss of the chromosome with the normal allele and a duplication of the chromosome with the defective allele; (2) recombination in the somatic cell by sister chromatid exchange, resulting in the loss of the normal allele and the presence of two chromosomes, each with the defective allele; (3) deletions or point mutations within the normal allele, leading to a loss of function and a homozygous defective cell; or (4) methylation of the promoter sequences of the normal gene, resulting in a loss of transcription and silencing of the second allele.

Examples of Human Tumor Suppressor Genes

Increasing numbers of tumor suppressor genes are being identified and shown to be associated with specific types of cancer. A list of some of the known tumor suppressor genes is given in Figure **C**. Tumor suppressor genes are found throughout the human genome and involve many different genes. Several cancers that have been shown to be associated with mutation in a tumor suppressor gene, leading to the loss of a tumor suppressor protein, include retinoblastoma (RB), Li-Fraumeni syndrome, early-onset breast and ovarian cancer (BRCA), familial adenomatous polyposis (FAP), and neurofibromatosis 1 and 2 (NF-1, NF-2).

For more information see Hoffee PA: *IMS: Medical Molecular Genetics*. Fence Creek, pp. 178–184.

23 Tumor Suppressor Protein pRb Regulates Cell Growth

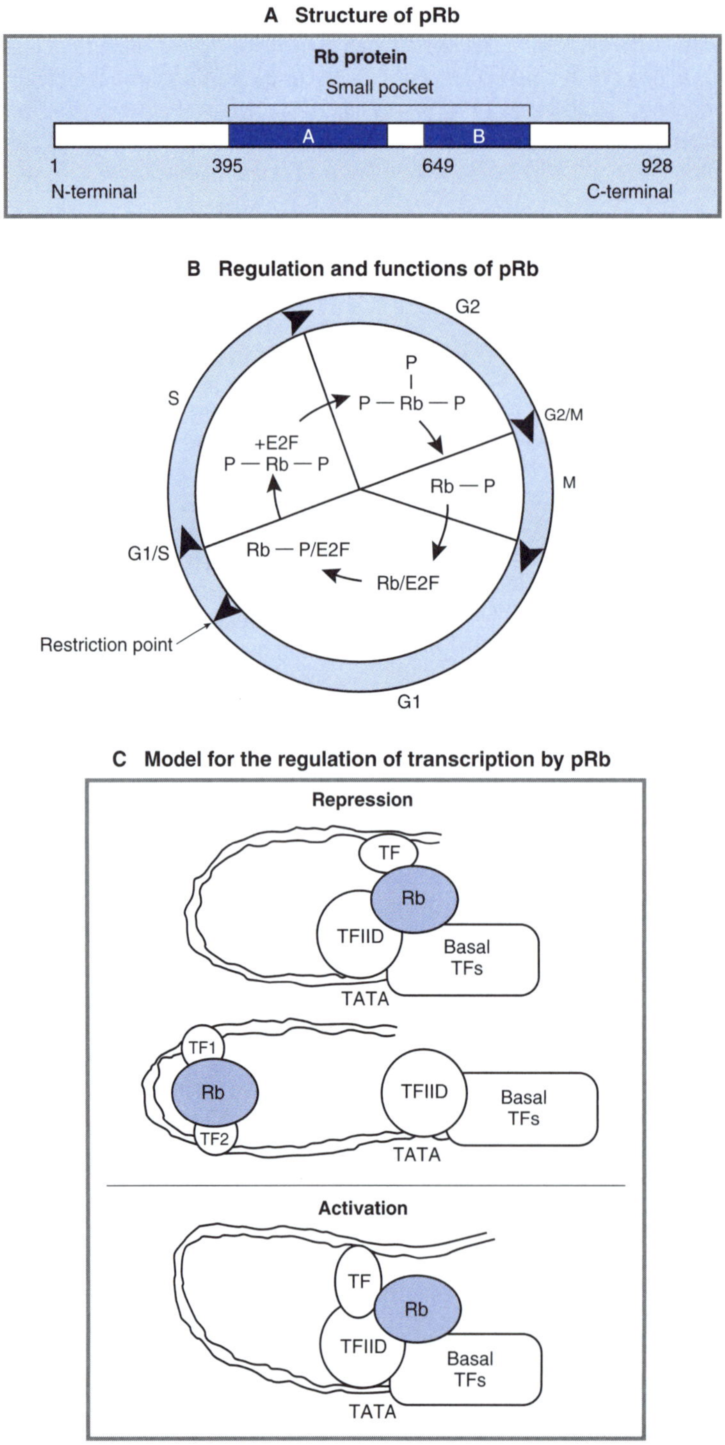

OVERVIEW

The retinoblastoma (RB) gene spans over 200 kb of DNA and codes for a protein (pRb) with a molecular weight of 105 to 110 kD. pRb is expressed in the nucleus of all cells; it regulates several important processes including cell cycle progression, cell differentiation, and apoptosis; and is known to act at the level of gene transcription by forming protein–protein complexes with both upstream TFs and basal TFs. This interaction can result in either positive or negative stimulation of expression of the genes coding for proteins that regulate progression through the cell cycle. Most mutations that occur in the *RB* gene lead to an alteration or a loss of the protein domain that interacts with other TFs. These mutations are the cause of retinoblastoma, a disease characterized by eye tumors in young children.

Structure of pRb

Transcriptional regulation is the main function of pRb, and it occurs by the interaction of pRb with other TFs bound to core promoter and upstream promoter elements. pRb itself does not bind directly to DNA and does not contain a DNA-binding domain (Figure **A**). A major portion of the protein structure—the A and B domain—is necessary for the interaction of pRb with other TFs. TFs that bind to pRb include E2F, a protein regulator of genes involved in DNA replication, and MYC and ELF-1, which are sequence-specific binding proteins that regulate transcription initiation. pRb also associates with the SWI/SNF complex of RNA polymerase II, the complex that is required for denaturing chromatin structure. The interaction of pRb with other TFs can result in either a repression or a stimulation of gene transcription. For example, binding of pRb to E2F represses transcription, whereas binding of pRb to MYC activates transcription.

Regulation and Functions of pRb

The functioning of pRb is controlled by its phosphorylation state, which varies as a cell passes through the phases of the cell cycle (Figure **B**). Cells in the G1 phase of the cell cycle contain the unphosphorylated form of pRb, which begins to be phosphorylated prior to the G1/S border. As the cells pass the G1/S border and enter into the S and G2 phases of the cell cycle, pRb undergoes further phosphorylation. As cells leave the M phase and proceed to a new G1 phase and the beginning of a new cell cycle, dephosphorylation of pRb occurs. When pRb is in its unphosphorylated state (i.e., as in G1), it is thought to be in its growth-restricting state. Phosphorylation of pRb, which leads to a loss of its negative functions, is carried out by specific cyclin-dependent kinases that function at various phases of the cell cycle (see Chapter 5). The interaction of pRb with E2F is an important control process in regulating progression through the cell cycle. When E2F is bound to the unphosphorylated form of pRb, the complex is still capable of binding to DNA but is no longer able to activate transcription and functions as a repressor that silences certain genes. When a cell is presented with growth-stimulating factors in G1, phosphorylation of pRb by cdk4–cyclin D or cdk2–cyclin E occurs, releasing E2F from pRb. With pRb no longer bound and not blocking the transactivation domain of E2F, E2F can now activate the transcription of genes necessary for the cell to pass the G1/S border and enter into the S phase. Once the cells have passed the G1/S border, they are committed to cell division.

Model for the Regulation of Transcription by pRb

pRb can regulate some genes by repressing their transcription. In one model (Figure **C**, upper panel), pRb blocks the binding of the TF to the basal transcription complex, thus preventing activation of transcription. In another model (Figure **C**, middle panel), pRb binds to two upstream TFs, TF1 and TF2, and prevents their interaction with the basal transcription complex and their activation of the complex. pRb also can regulate transcription in a positive way (Figure **C**, bottom panel) by stimulating transcription mediated by a TF such as SP1. pRb can interact with TAFIID 250, a protein that is part of the basal transcription complex, and the TF at the same time in such a way that the basal transcription complex is stabilized and activated to carry out transcription. pRb may stimulate transcription by Sp1 by its interaction with TAFIID 250 and some of the other TAFs, such as TAFIID 110, which is known to interact with Sp1. This interaction between the basal transcription complex, the upstream TFs, and pRb provides the cell with a tight connection between transcription initiation and cell cycle progression.

Clinical Significance

When mutations occur in both *RB* alleles, pRb is missing or has an altered function, and the regulation of transcription as well as cell cycle regulation is lost. The defective cell continues to progress through the cell cycle no longer subject to normal controls and is destined to become a tumor cell. The development of eye tumors (i.e., retinoblastoma) in children is due to the loss of pRb function. Two forms of retinoblastoma occur: the familial form, in which one defective allele of the *RB* gene is passed through the germline with the second defective allele arising in somatic cells, and the sporadic form, in which both defective alleles arise at the somatic cell level. Approximately 40% of retinoblastoma patients have the familial form and a family history of the disease. In these individuals, tumors appear at an average age of 14 months and are usually bilateral; that is, tumors are present in both eyes. The remaining 60% of retinoblastoma cases occur sporadically with no history of the disease in the family. In these individuals, the tumors appear later (approximately 30 months of age) and are usually present in only one eye.

For more information see Hoffee PA: *IMS: Medical Molecular Genetics*. Fence Creek, pp. 179–183.

24 Complex Transcriptional Units and Translational Control

A Promoter selection in the α-amylase gene

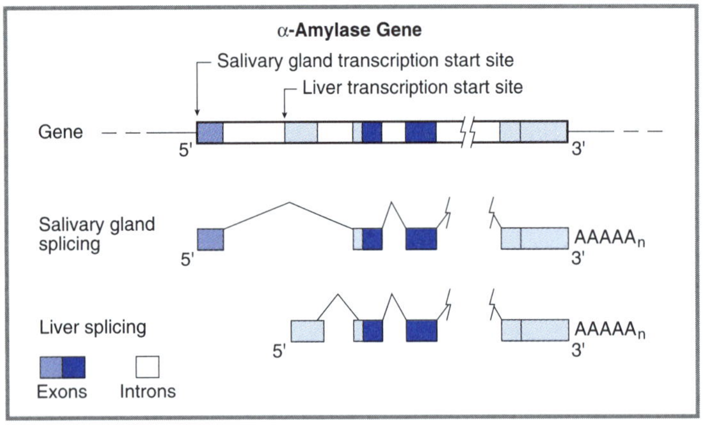

B Alternative polyadenylation sites in the immunoglobulin genes

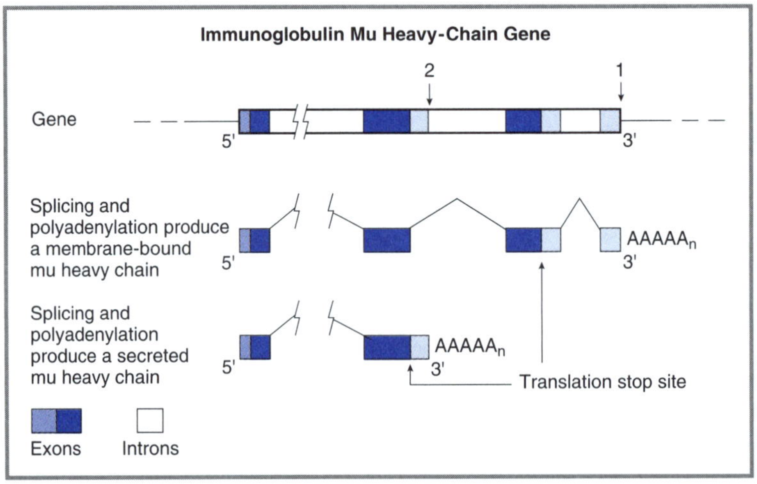

C Control of mRNA stability and mRNA translation

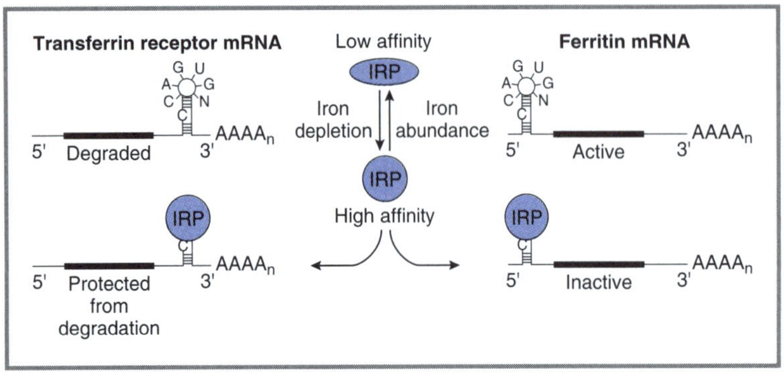

OVERVIEW

In addition to regulating gene expression at the level of transcriptional initiation, cells have other ways of controlling the expression of genes. They include promoter selection, alternative splicing of exons or differential polyadenylation of the precursor mRNA, controlling the stability or the rate of degradation of the mature mRNA, and controlling the process of translation.

Promoter Selection in the α-Amylase Gene

Located within the α-amylase gene are two promoter sites that control the expression of this gene in a tissue-specific manner and provide a way for a cell to produce the α-amylase protein at high levels, at low levels, or not at all (Figure **A**). The α-amylase gene contains two 5′ untranslated exons (*gray boxes*) and two translated exons (*black boxes*), which contain the coding information that determines the amino acid sequence of the final protein product. In the salivary gland, TFs recognize the first promoter, located at the *arrow*. This promoter is a strong promoter and results in high levels of mRNA being produced in the salivary gland. In liver cells, TFs recognize the second and weaker promoter just 5′ to the second untranslated exon, indicated by the *arrow*. During the splicing process, the 5′ untranslated exons in both cases are spliced to the first exon containing the amino acid sequence information, resulting in a different mRNA being produced with respect to the 5′ untranslated region. The amino acid sequence of α-amylase is the same in both tissues; only the level of the protein differs.

Alternative Polyadenylation Sites in the Immunoglobulin Genes

Immunoglobulin M (IgM) is an antibody molecule that can occur either as a membrane-bound receptor for antigen or as a secreted form of the protein, depending on the state of development of B, or antibody-producing, cells. This differential production of a membrane or secreted form of IgM is dependent on the structure of the mu heavy chain that is part of the antibody molecule. The membrane form of IgM contains a mu heavy chain that has a carboxyl-terminal amino acid sequence rich in hydrophobic amino acids, facilitating its interaction and binding to the cellular membrane. The secreted form of the antibody contains a mu heavy chain that lacks this carboxyl-terminal amino acid sequence and thus is no longer capable of binding to the membrane. During B-cell development, a mechanism operates at the level of polyadenylation of the RNA that determines which mu chain will be synthesized. Figure **B** is a simplified schematic of the actual mu gene with only some of the exons represented. In cells in which the membrane-bound form of the antibody is produced, a polyadenylation signal present at the distal 3′ end of the gene, indicated by *arrow 1*, determines the site of cleavage and polyadenylation of the RNA. During the splicing process, all of the exons in this RNA, including the 3′ exon coding for the hydrophobic amino acids found at the carboxyl-terminal end of the membrane-bound form of the mu heavy chain, are spliced together to produce the mature mRNA. In cells in which the secreted form of the antibody is produced, the upstream polyadenylation signal, *arrow 2*, is recognized and determines the site of cleavage and polyadenylation of the RNA found in mature B cells. The mRNA produced after splicing is missing the exons located 3′ to this polyadenylation signal, resulting in the production of a secreted form of the mu heavy chain that no longer contains a hydrophobic tail.

Control of mRNA Stability and mRNA Translation

The amount of a protein present in a cell is directly related to the amount of mature mRNA present. This dependent relationship offers another level at which a cell can regulate protein concentrations. The steady-state level of a mRNA species is a balance between its rate of synthesis and its rate of decay, or half-life. The half-life of mRNAs can vary in human cells from 30 minutes to hours and is controlled to some extent by the length of the polyA sequence attached at the 3′ end of the message. Control at the level of transcription determines the rate of synthesis of a mRNA, whereas control at the level of stability determines the rate of decay. An example in which the level of a protein is determined by altering the rate of decay of the mRNA is the synthesis of the transferrin receptor protein, which is responsible for internalizing free iron carried in the blood by the carrier protein transferrin (Figure **C**).

Located at the 3′ end of the transferrin receptor mRNA is a stem loop structure, the iron response element, that binds an iron regulatory protein (IRP), when the cell is depleted of iron. Binding of the IRP to the 3′ end of the transferrin receptor mRNA protects the mRNA from degradation and results in an increase in the level of the transferrin receptor mRNA and a corresponding increase in the level of the transferrin receptor protein. In the presence of an increased level of the transferrin receptor protein, more iron can be internalized, thus raising the intracellular iron concentration. When the concentration of iron is in excess, it must be bound by the intracellular protein ferritin, or it will be toxic to the cells. In the presence of high concentrations of iron, the level of ferritin increases without a corresponding increase in the synthesis of ferritin mRNA.

Located in the 5′ UTR of the ferritin mRNA molecule is a stem-loop structure that binds IRP when iron is depleted in the cell. Binding of the IRP to the 5′ end of the ferritin mRNA blocks the translation of this mRNA and results in a decrease in the level of ferritin protein. When iron is in abundance, the ferritin mRNA no longer binds IRP and actively translates ferritin protein. At the same time, iron abundance will prevent IRP from binding to the 3′ end of the transferrin receptor mRNA, and the mRNA will be degraded, reducing the level of the transferrin receptor protein.

For more information see Hoffee PA: *IMS: Medical Molecular Genetics*. Fence Creek, pp. 184–188.

PART I: QUESTIONS

Directions: For each of the following questions, choose the **one best** answer.

Questions 1–6

1. Both DNA and RNA contain

(A) thymine

(B) ribose

(C) phosphodiester bonds

(D) uracil

(E) deoxyribose

2. Double-stranded DNA contains 30 mol percent thymine. What is the mol percent cytosine?

(A) 20

(B) 30

(C) 40

(D) 60

(E) Insufficient information to determine

3. Which of the following statements describes the replication of human chromosomes?

(A) Replication of the leading strand requires only a single DNA primer.

(B) Replication proceeds bidirectionally from multiple points on the chromosomes.

(C) Both DNA strands are synthesized in a 3′ to 5′ direction.

(D) Replication requires a single type of DNA polymerase.

(E) Replication initiates on one chromosome at a time.

4. Which of the following events occurs during meiosis but not during mitosis?

(A) Chromosome alignment at the metaphase plate

(B) The formation of spindles

(C) Division of the centromeres

(D) Condensation of the chromosomes

(E) Pairing of homologous chromosomes

5. Which of the following general transcription factors contains the TATA-binding protein (TBP)?

(A) TFIIA

(B) TFIIB

(C) TFIID

(D) TFIIE

(E) TFIIH

6. Which of the following proteins has the ability to disrupt chromatin structure in the presence of ATP?

(A) TFIID

(B) AP1

(C) SWI/SNF

(D) STATS

Directions: The groups of questions below consist of lettered choices followed by several numbered items. For each numbered item, select the appropriate lettered option with which it is most closely associated. Each lettered option may be used once, more than once, or not at all.

Questions 7–10

Match each of the following definitions with the enzymes listed below.

(A) DNA helicase

(B) DNA polymerase-δ

(C) DNA topoisomerase

(D) DNA ligase

(E) Ribonuclease H1

(F) DNA polymerase-β

(G) 3′ to 5′ exonuclease

7. This enzyme is necessary for relaxing the tension in a replicating DNA molecule and catalyzes the breaking and resealing of phosphodiester bonds.

8. This enzyme breaks hydrogen bonds to facilitate the unwinding of DNA.

9. This enzyme catalyzes the removal of mispaired bases formed by DNA polymerase during DNA replication.

10. This enzyme seals the gap between adjacent nucleotides in DNA during DNA replication and also during DNA repair.

Questions 11–13

Match each of the following descriptions with the term that it best fits.

(A) Nucleosome

(B) Telomere

(C) Centromere

(D) Histone

(E) Chromatid

11. This element ensures correct segregation of homologous chromosomes during meiosis and mitosis.

12. This element occurs as repetitive sequences located at the ends of chromosomes and functions to maintain the integrity of the chromosome.

13. This element is composed of DNA and basic proteins and facilitates the compacting of DNA within the nucleus.

Questions 14–16
Match the following descriptions with the term it best fits.

(A) G banding

(B) Karyotype

(C) Metaphase spread

(D) FISH

(E) p arm

14. This procedure is a standard procedure for identifying metaphase chromosomes in a human karyotype.

15. This is the standard designation for part of a human chromosome.

16. This technique uses fluorescent dyes to localize a specific gene sequence within a human chromosome.

Questions 17–19
Match the following functions that occur in the mammalian cell cycle with the correct phase.

(A) G0 phase

(B) G1 phase

(C) G2 phase

(D) S phase

(E) M phase

17. During this phase of the cell cycle, the total complement of DNA must be replicated.

18. During this phase of the cell cycle, the chromosomes are visible under a light microscope as unique identifiable structures.

19. Cells not undergoing cell division exist in this phase.

Questions 20–21
Match the following functions that occur during mitosis with the correct phase.

(A) Prophase

(B) Prometaphase

(C) Metaphase

(D) Anaphase

(E) Telophase

20. During this phase of mitosis, a protein binds the sister chromatids together and prevents them from separating until the appropriate signal is given.

21. During this phase of mitosis, the sister chromatids are pulled apart, and one chromatid from each chromosome moves to the opposite poles of the cell.

Questions 22–25
A diagram of a typical human gene containing N exons is shown below. The *arrow* designates the transcription start site at the 5' end of the gene. Exons are indicated by the *darker shaded boxes* and numbered as 1, 2, up to N. The lighter shaded boxes indicate the 5' and 3' untranslated regions. Introns are indicated by the *open boxes*. The *vertical lines* designate regions within the gene in which a small 4-bp deletion causes an inherited disease. Match the following mutation, indicated as A, B, C, or D, with its most likely effect on the amount or structure of the protein coded for by the gene.

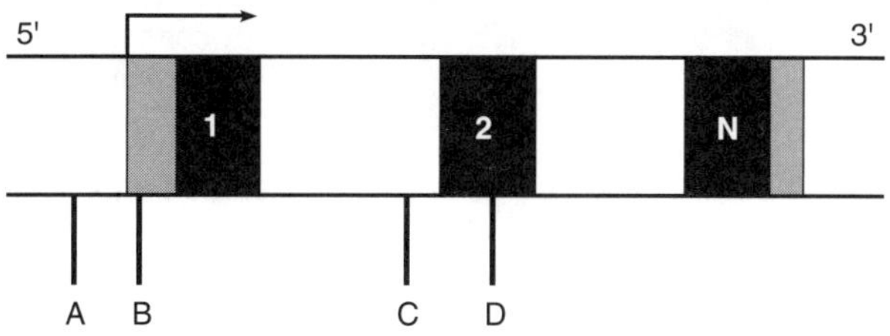

22. This mutation prevents transcription of the gene.

23. This mutation causes a change in the amino acid sequence of the protein.

24. This mutation blocks the initiation of translation of the mRNA.

25. The mutation alters the production of a functional mRNA from the primary RNA transcript.

Questions 26–28
Match the following events with the appropriate term.

(A) Capping

(B) AAUAAA

(C) Spliceosome

(D) Lariat structure

(E) CAAT

26. These are small ribonuclear bodies made up of RNA and protein that carry out splicing of the precursor RNA.

27. This is the first step in the processing of precursor RNA to mRNA and uses a nucleoside called 7-methylguanosine.

28. This signal occurs at the 3′ end of the primary RNA transcript and signals the cleaving of the RNA by a nuclease.

Questions 29–31
Match each of the following descriptions with the most appropriate term.

(A) AUG

(B) UAG

(C) Anticodon

(D) Codon

(E) tRNA

29. This is the signal in all mRNA molecules for translation initiation to occur.

30. This signal found on all tRNA molecules recognizes a specific triplet of bases in mRNA.

31. This signal in mRNA is not recognized by a tRNA molecule and results in termination of protein synthesis.

Questions 32–34

The amino acid sequence of the carboxyl-end of a normal protein is the following:

Asn-Thr-Pro-Ser-Ile-Glu-Lys-Pro-COOH

In the following three cases, a mutant form of the protein has been identified with the designated amino acid sequence. Match the following explanations for the differences seen between the amino acid sequence of each mutant protein and that of the normal protein (the genetic code is given in Chapter 11).

(A) A single amino acid change caused by a single base-pair substitution

(B) A single amino acid change caused by a single base-pair deletion

(C) Premature termination of translation caused by a single base-pair substitution

(D) A frameshift caused by a single base-pair addition

(E) A frameshift caused by a single base-pair deletion

32. Mutant 1 produces a protein with this sequence: Asn-Thr-Pro-Ser-Ile-COOH.

33. Mutant 2 produces a protein with this sequence: Asn-Thr-Pro-Cys-Ile-Glu-Lys-Pro-COOH.

34. Mutant 3 produces a protein with this sequence: Asn-Thr-Pro-Ser-Ile-Asp-Glu-Ala-Leu-Val-Ala-Arg-Gly-Val-COOH

Questions 35–37

DNA repair is a very important process in all living cells and is responsible for removing mistakes or damaged bases from DNA. The four main repair systems, listed as A–D, each serve a specific role in human cells to prevent mutations leading to disease from occurring. Match the following roles with the specific system.

(A) Proofreading by the 3′ to 5′ exonuclease of DNA polymerase

(B) The methyl-directed mismatch repair system

(C) The nucleotide-excision repair system

(D) The base-excision repair system

35. This system removes mispairs from the DNA immediately after DNA replication.

36. This system recognizes damaged or altered bases in DNA.

37. This system repairs bulky distortions in the DNA and is particularly important for the repair of damage caused by UV light.

Questions 38–40

Match each of the following definitions to the term it best describes.

(A) mRNA

(B) tRNA

(C) rRNA

(D) hnRNA

(E) snRNA

38. This RNA molecule carries amino acids to the ribosomes during protein synthesis.

39. This RNA molecule carries information for the amino acid sequence of a protein to the protein synthesis machinery.

40. This RNA molecule is the immediate product of transcription and the precursor to mRNA.

PART I: ANSWERS AND EXPLANATIONS

1. **The answer is C.**

The nucleotides in both DNA and RNA molecules are held together by phosphodiester bonds between the 3′ hydroxyl group and the 5′ phosphate group of adjacent nucleotides to form a single chain. DNA contains thymine and deoxyribose, whereas RNA contains uracil and ribose.

2. **The answer is A.**

Double-stranded DNA has an equal number of A and T residues and an equal number of G and C residues. If T equals 30%, A must equal 30%, making A-T equal to 60%. The remainder is 40%, with 20% for C and 20% for G.

3. **The answer is B.**

DNA replication is bidirectional and initiates at multiple sites on each chromosome to allow the total genome to be replicated during the S phase. The leading strand requires a single RNA primer, and both the leading and lagging strands are synthesized in a 5′ to 3′ direction. Both DNA polymerase-α and DNA polymerase-δ are required for DNA replication. Initiation of DNA replication occurs at the same time on all human chromosomes.

4. **The answer is E.**

Homologous chromosomes pair during prophase I of meiosis I to give a bivalent made up of four chromatids. The paired (synapsed) chromosomes now are in position to undergo recombination, allowing new combinations of genes to be transferred to the gametes in the form of recombinant chromosomes. All of the other processes occur both in mitosis and in meiosis.

5. **The answer is C.**

TFIID is a complex made up of approximately ten different subunits, one of which is TBP, the protein that binds to the TATA sequence in promoters of human genes. In addition, TFIID contains proteins known as TBP-associated factors, which vary in molecular weight from 10 to 25 kD.

6. **The answer is C.**

The SWI/SNF protein complex is made up of at least ten proteins, including a subunit with ATPase activity. In the presence of ATP, the complex disrupts the nucleosome array and facilitates transcription of DNA previously unavailable to the transcription complex. TFIID is a

GTF that is bound by way of the TBP to the TATA sequence. AP1 is an inducible TF that binds RE sequences. STATs, signal transducers and activators of transcription, are TFs that are important in the response of a cell to cytokines.

7–10. **The answers are: 7-C, 8-A, 9-G, 10-D.**

DNA topoisomerase acts downstream of the replicating fork to release the tension on the supercoiled DNA. It does so by breaking and resealing phosphodiester bonds. DNA helicase acts at the replication fork, opening the double-stranded DNA to form single strands and allowing the DNA polymerase to move down the strand. It also serves the same unwinding function during DNA repair and transcription. The 3′ to 5′ exonuclease, an activity associated with DNA polymerases involved in DNA replication, recognizes a mispaired base during DNA replication and removes the mispaired base, allowing DNA polymerase a second chance to incorporate the correct nucleotide into the growing DNA strand. DNA ligase seals the 3′ terminal hydroxyl and the 5′ terminal phosphate group of adjacent nucleotides in DNA. These gaps occur during DNA replication and during the DNA repair processes.

11–13. **The answers are: 11-C, 12-B, 13-A.**

Centromeres are the constrictions on chromosomes that separate the long and short arms and are essential for correct segregation of the homologous chromosomes in meiosis and mitosis. Telomeres are made up of the repetitive DNA sequence, TTAGGG, and are located on both ends of each chromatid. They are essential for maintaining stability of the chromosomes and assuring the proper replication of the ends of the chromosomes. Nucleosomes are structures found within the nucleus and are made up of DNA and histones. Each nucleosome has eight histone proteins with 146 bp of DNA wrapped around them to form a defined structure that allows the DNA to be compacted in the nucleus.

14–16. **The answers are: 14-A, 15-E, 16-D.**

G banding is a procedure that combines Giemsa stain with trypsin treatment to produce light and dark bands on human metaphase chromosomes. The location of bands has been standardized for each human chromosome and allows the precise identification of each human chromosome. With standardized nomenclature, a karyotype that deviates from the normal can be clearly identified. Each human chromosome is divided by the position of the centromere into the p, or short, arm and the q, or long, arm. Therefore, the positions of genes within a chromosome are named according to their location within a band on the p or q arm (i.e., 2p16 means a location on chromosome 2, the p arm at region 1, band 6). FISH is a procedure in which a DNA probe is labeled with a fluorescent dye and then used to localize gene sequences within specific chromosomes.

17–19. **The answers are: 17-D, 18-E, 19-A.**

During the S, or synthesis, phase of the cell cycle, all of the chromosomes must be totally replicated, but only once. When this phase is completed each chromosome consists of two identical chromatids. The M, or mitotic phase of the cell cycle is the time at which metaphase chromosomes are visible. Therefore, banding procedures used to identify chromosomes are carried out on cells in the M phase. Additionally, during the M phase, the chromosomes that were duplicated during the S phase are equally distributed to two daughter cells. Cells not undergoing active cell divisions, such as neurons, are removed from the cell cycle and exist in the G0 phase. If cells in the G0 phase are stimulated to grow, they enter into the G1 phase and the cell cycle begins.

20 and 21. **The answers are: 20-C, 21-D.**

During metaphase, all of the chromosomes are aligned at the metaphase plate. The two sister chromatids are held together by the binding of a protein (an inhibitor of sister chromatid separation [ISS]) until all the chromosomes are properly aligned. With the degradation of the ISS protein, anaphase begins. Anaphase is the phase during which the sister chromatids move to opposite poles. Accurate distribution of each chromosome to each daughter cell is crucial to prevent chromosomal abnormalities such as monosomies or trisomies.

22–25. **The answers are: 22-A, 23-D, 24-B, 25-C.**

A mutation at position A most likely changes the base-pair sequence within the promoter of the gene, resulting in a lack or decrease in the transcription of the gene by interfering with the binding of RNA polymerase II or various TFs. A 4-bp deletion at position D alters the amino acid sequence of the protein because of a shift in the reading frame. A mutation at position B in the 5′ untranslated region blocks the translation initiation signals and interferes with mRNA translation. A mutation at position C within an intron can change the splicing signals and result in a defect in the splicing of the precursor RNA to a mature mRNA.

26–28. **The answers are: 26-C, 27-A, 28-B.**

Spliceosomes are small ribonuclear bodies that carry out RNA splicing. They function by recognizing the splice donor and splice acceptor sites at the beginning and end of each intron. The RNA components of the spliceosomes are the snRNAs. Capping is the first step in the processing of precursor RNA and involves the attachment to the 5′ end of the RNA of the nucleoside 7-methylguanosine. The sequence AAUAAA in the 3′UTR of the precursor RNA is known as the poly A addition signal. This sequence is recognized by a nuclease, resulting in cleavage of the RNA and the addition of adenosine residues to the end of the RNA (poly A tails).

29–31. **The answers are: 29-A, 30-C, 31-B.**

AUG, the codon for methionine, is the signal for translation initiation. During the translation process, the 40S ribosome subunit scans the mRNA until it reaches the first AUG codon. At this point, the 60S subunit binds the complex, and the initiation step is completed. The anticodon is a three-base sequence found on all tRNA molecules and is complementary to a specific codon found in the mRNA. The codon in the mRNA is recognized by the anticodon of the tRNA, allowing the tRNA to attach to the mRNA by base complementarity. The UAG codon is one of the three termination codons, the other two being UAA and UGA, found in mRNA molecules. No tRNA molecules have a complementary anticodon for these three codons; therefore, no amino acid is added to the growing polypeptide chain and protein synthesis terminates.

32–34. **The answers are: 32-C, 33-A, 34-D.**

Mutant 1 has a single base-pair substitution, in which the first G of the codon for glutamic acid (GAA or GAG) has been changed to a U to give one of the stop codons (UAA or UAG). Mutant 2 has a single base-pair substitution in which the first A of the codon for Ser (AGU or AGC) has been changed to U to give the codon for Cys (UGU or UGC). Mutant 3 has a single base-pair addition, leading to a frameshift mutation. The normal sequence for Glu-Lys-Pro-COOH would be GAG-AAG-CCC-UAG (STOP)-COOH. The mutant sequence resulting from a single base-pair addition to give Asp-Glu-Ala-Leu-Val would be GA**<u>C</u>**-GAA-GCC-CUA-GUG, etc. The addition of a C residue (underlined and in bold) shifts the reading frame and changes all the amino acids in the protein from the point of insertion. The shift changes the normal stop codon, UAG, to a normal sense codon, CUA, making the protein longer than the normal.

35–37. **The answers are: 35-B, 36-D, 37-C.**

The methyl-directed mismatch repair system recognizes a mispair in the DNA that has occurred from a mistake by DNA polymerase. The repair system recognizes the incorrect base on the newly replicated DNA strand because that strand is not yet methylated. Correction takes place specifically on the nonmethylated newly synthesized DNA strand. Defects in this repair system can lead to the human disease HNPCC. The base-excision repair system removes altered or damaged bases and is of particular importance in removing uracil residues from DNA. Uracil, which is not a normal component of DNA, arises in the DNA by spontaneous deamination of cytosine residues. If not removed, the original G-C pair will become an A-T pair. The nucleotide-excision repair system is important in removing the thymine–thymine (T–T) dimers that arise after exposure to sunlight and UV irradiation. Defects in the nucleotide excision repair system lead to the autosomal recessive disorder xeroderma pigmentosum.

38–40. **The answers are: 38-B, 39-A, 40-D.**

tRNA molecules bind specific amino acids and participate in the process of protein synthesis by binding to the mRNA by way of the anticodon and then placing the correct amino acid in the growing polypeptide chain. mRNA, once processed, is made up of only exon sequences, some of which contain the sequence information for amino acids. The mRNA then moves to the ribosomes, where it is translated into protein. hnRNA is the immediate product of transcription by RNA polymerase II and contains both intron and exon sequences. This RNA is then capped at the 5′ end, polyadenylated at the 3′ end and the introns spliced out to produce the final mature mRNA molecule.

PART II
Mendelian Inherited Diseases

25 Mendelian Inheritance

A Standard pedigree symbols, definitions, and abbreviations

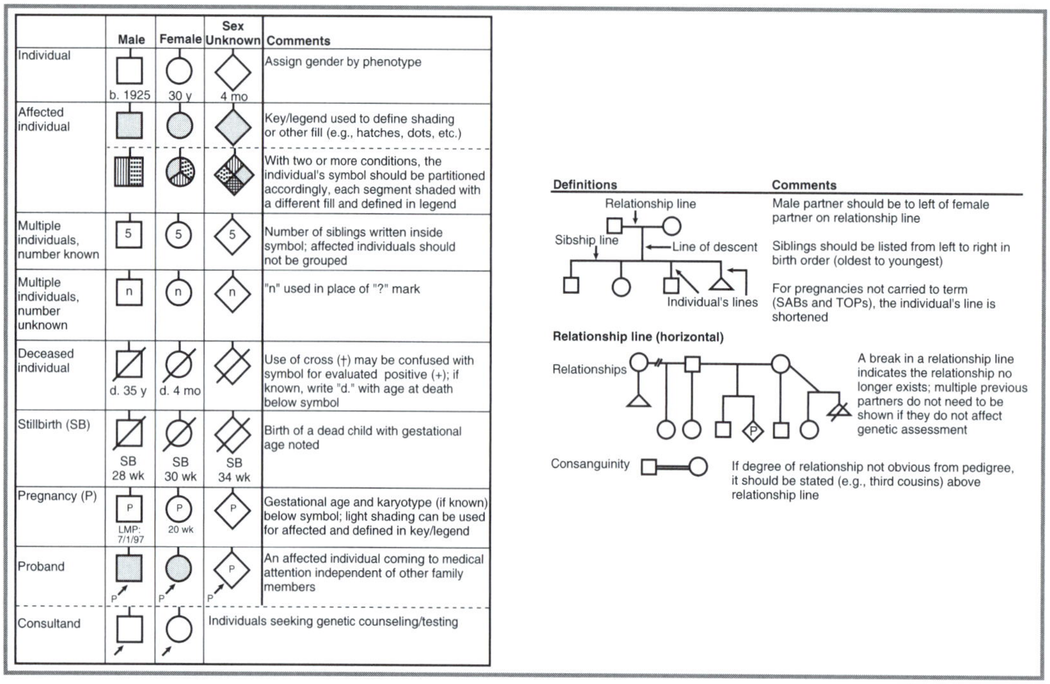

B Segregation of alleles

		Parent one (Genotype AB)	
	Gametes	A	B
Parent two (genotype CD)	C	AC	BC
	D	AD	BD

Note: Shaded area signifies the possible offspring from the mating of an *AB* parent and a *CD* parent.

C Dominant-recessive alleles

	Phenotype if:	
Genotype	***A Allele dominant/ B Allele recessive***	***A Allele recessive/ B Allele dominant***
AA	Affected	Affected
AB	Affected	Unaffected
BB	Unaffected	Unaffected

D Independent assortment of alleles

		Parent one (Genotype BC; 12)			
	Gametes	B, 1	B, 2	C, 1	C, 2
Parent two (genotype AA; 12)	A, 1	AB; 11	AB; 12	AC; 11	AC; 12
	A, 2	AB; 12	AB; 22	AC; 12	AC; 22

Note: Shaded area signifies all possible offspring from a mating of *(AA; 12)* x *(BC; 12)* parents, illustrating independent assortment of the alleles at the two loci.

OVERVIEW

In 1865, Gregor Mendel reported the basic principles for the inheritance of alleles at a single genetic locus (i.e., monogenic). These inheritance patterns, referred to as mendelian inheritance, fall into four classes depending on the location of the genetic locus (i.e., on an autosome or on a sex chromosome) and the dominant-recessive pattern of the affected allele. Mendelian inheritance patterns for monogenic diseases are referred to as autosomal recessive, autosomal dominant, sex-linked recessive, or sex-linked dominant. A characteristic pedigree diagram is observed for each pattern of inheritance.

Standard Pedigree Symbols, Definitions, and Abbreviations

Pedigree diagrams are the most common way to depict inheritance patterns in families that exhibit inherited diseases. These pedigrees are developed by taking family histories, observing the phenotype of the individuals, examining medical records, and, more recently, performing DNA analysis. Important aspects of the pedigree diagram (Figure **A**) include the use of a box (male) or circle (female) for the assignment of gender, the relationship lines for married individuals and their offspring, the listing of each generation (I, II, III, etc.) known, the listing of siblings from left to right in order of their birth (noted as 1, 2, 3, etc.), shading of the symbol for an affected individual, the use of a slash to denote a deceased individual, the notation of an arrow to denote the proband, and the use of a female symbol with a dot to represent the carrier status for an X-linked recessive disorder.

Segregation of Alleles

Three basic principles arose from Mendel's work. (1) Characteristics are inherited as units and not blended. Although characteristics might not be expressed in the offspring of the first generation, they could appear unchanged in the second generation. (2) The principle of segregation of alleles states that each parent possesses alleles in pairs, and only one allele of each parent is transmitted or segregated to the offspring. The segregation of alleles that are not blended allows inheritance to be traced from one generation to the next. (3) The law of independent assortment states that different pairs of alleles assort randomly in the gametes (see below).

The Punnett square, developed by Reginald Punnett nearly 50 years after Mendel's original work, is used to demonstrate the segregation of alleles to the gametes and allows the prediction of the relative proportions of the offspring genotypes (Figure **B**). In mating between parent *AB* and parent *CD*, each parent produces two gametes (i.e., parent *AB* has an *A* and a *B* gamete, and parent *CD* has a *C* and a *D* gamete). The four possible offspring are found in equal proportions: ¼ *AC*, ¼ *AD*, ¼ *BC*, and ¼ *BD*.

Dominant-Recessive Alleles

Although Mendel's work clearly demonstrated that alleles are not blended, it also showed that one allele is capable of masking another allele (i.e., one allele is dominant and one allele is recessive). For a dominant allele, the phenotype of a heterozygote (an individual with two forms of an allele) is the same as the phenotype of a homozygote (an individual in whom both alleles are the same). In the case of dominance, only one copy of an allele is necessary to express the phenotype. For a recessive allele, two copies of an allele are necessary to express the phenotype. Consider the example shown in Figure **C** of a two-allele autosomal genetic locus for a disorder with an affected and an unaffected phenotype, *A* representing the affected and *B* representing the unaffected phenotype. In one case, *A* is dominant and *B* is recessive; therefore, both *AA* and *AB* are affected, and only *BB* is unaffected. In the other example, where *A* is recessive and *B* is dominant, only *AA* is affected, and both *AB* and *BB* remain unaffected.

Independent Assortment of Alleles

In Mendel's third principle, the law of independent assortment states that different pairs of alleles assort randomly and independently of one another in the gametes. In particular, alleles that are present at different genetic loci are transmitted independently to the gametes. The understanding of independent assortment is an important concept in the study of genetic linkage (see Chapter 48). The Punnett square in Figure **D** illustrates the concept of independent assortment for a mating of two parents with two different loci (parent *AA, 1, 2*; and parent *BC, 1, 2*). With two loci segregating independently, six different combinations are possible in the offspring.

Other Modes of Inheritance

Most of the information available today on the relationship of defective genes to human diseases is concerned with monogenic inheritance, in which the phenotype is due to the inheritance of alleles at a single genetic loci. In addition to monogenic inheritance, four other modes of inheritance are known: oligogenic, wherein the phenotype is a result of the interaction of a few genetic loci; polygenic, or multilocus, wherein the phenotype is due to many genetic loci; multifactorial, wherein the phenotype is due to the interaction of many genetic loci and environmental factors; and mixed, wherein the phenotype is due to a single genetic locus plus multifactorial modification.

Clinical Significance

Inherited human diseases are studied by examining family histories and developing standardized pedigree diagrams that can be interpreted with regard to the patterns of inheritance predicted by Mendel's laws. Each of the four major types of inheritance patterns for monogenic diseases (autosomal recessive, autosomal dominant, sex-linked recessive, and sex-linked dominant) has specific sets of characteristics that can be used to accurately determine the mode of inheritance of a particular disease. This information is essential in providing information to family members with regard to carrier status, prenatal diagnosis, and the potential for the transmission of a diseased allele to future offspring.

For more information see Hoffee PA: *IMS: Medical Molecular Genetics*. Fence Creek, pp. 237–243.

26 Autosomal Recessive Inheritance I

General Concepts

A Typical autosomal recessive pedigree

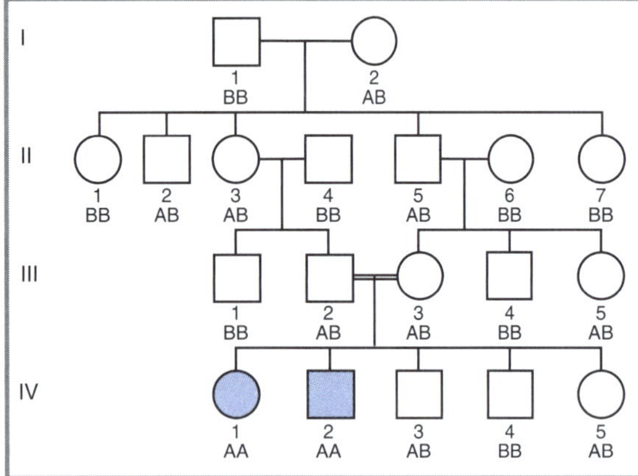

		Unaffected parent (AB)	
	Gametes	A	B
Unaffected parent (AB)	A	AA	AB
	B	AB	BB

Note: Shaded area signifies affected offspring genotypes.

B Autosomal recessive inherited disorders

Disorder	*Protein altered*	*Frequency*
Sickle cell anemia	β-Globin	1/650 blacks (in United States)
Cystic fibrosis	Transmembrane conductance regulator	1/2500 whites
Tay-Sachs disease	Hexosaminidase A	1/3000 Ashkenazi Jews
Gaucher disease	β-Glucosidase	1/6000 (1/600 Ashkenazi Jews)
Phenylketonuria	Phenylalanine hydroxylase	1/12,000
Severe combined immunodeficiency disease	Adenosine deaminase	Extremely rare

C Glycolipid degradation in the lysosomes

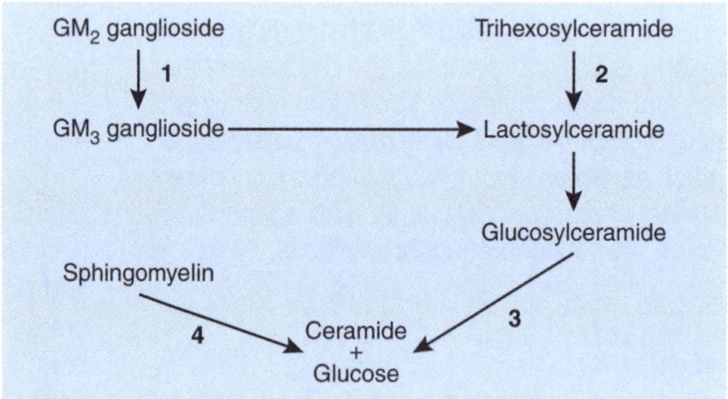

D Conversion of phenylalanine to tyrosine

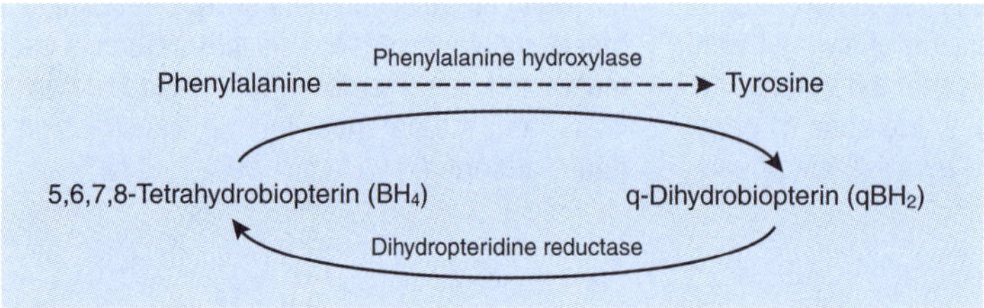

OVERVIEW

Autosomal recessive inheritance is generally attributed to loss-of-function mutations in which both copies of a patient's gene are dysfunctional. For most genes, one functional copy is sufficient to preclude clinical symptoms, although individuals with one inactive allele of a gene are carriers of an autosomal recessive disease. Any mutation that disables a gene, the RNA, or the protein in such a way that little or no functional product is produced can lead to a loss-of-function mutation. These mutations include deletions; insertions; chromosomal breaks within a gene; defects in transcription, RNA processing, or translation; missense or nonsense codons in the middle of the coding region of the gene; and mutations that affect the stability of the RNA or protein product. Because there are many ways of inactivating a gene or disabling a gene product, loss-of-function mutations are quite common and account for many of the known inherited diseases.

Autosomal Recessive Pedigrees

A typical autosomal recessive pedigree is shown in Figure **A**, and illustrates four main characteristics of diseases that require the presence of two defective alleles to produce an affected phenotype:

1. On the average, 25% of the offspring for a mating between two unaffected heterozygotes (*AB* × *AB*, with *A* as the affected allele) will be affected with the disease. Among the offspring, 25% will be *AA* and affected, 50% will be *AB* and unaffected carriers, and 25% will be *BB* and unaffected noncarriers of the disease.
2. Vertical transmission is characteristic of the pedigree; that is, the phenotype usually occurs in siblings and cousins rather than in parents and offspring of affected individuals. Additionally, one does not see affected individuals in each generation.
3. Because both parents of an affected individual must carry the defective allele, the chance of the disease occurring is increased if the parents are related and have at least one ancestor in common.
4. For autosomal recessive disorders, males and females are equally affected.

Autosomal Recessive Inherited Disorders

Many of the known monogenic disorders that have been studied at the molecular level show autosomal recessive inheritance patterns and are the result of loss-of-function mutations in known genes (Figure **B**). Many of these genes normally code for proteins involved in cellular metabolism, such as the enzymes involved in the metabolism of carbohydrates, nucleotides, amino acids, lipids, and organic acids. One of the most commonly inherited disorders of the beta-globin gene is sickle cell anemia, which results from a single amino acid change of a glutamic acid to a valine at position 6 of the protein. Individuals homozygous for this mutation are affected with sickle cell disease, which is often fatal. The disease is most common in the U.S. African American population, with a frequency of 1 in 650 live births. The clinical phenotype of a child with sickle cell anemia is not obvious at birth but begins to become apparent within the first 2 years of life. The hemoglobin (Hb) molecule in an individual with sickle cell disease is made up of two normal alpha chains and two mutant beta chains. Although the mutant Hb molecules can still bind oxygen, when they are present in deoxygenated blood they become less soluble than normal Hb and tend to form aggregates of rod-shaped polymers. The presence of these aggregates in the RBCs causes a distortion and a sickling of the cells. Once distorted, the RBCs can no longer enter into the capillaries, resulting in a blockage and a lack of blood flow. The RBCs, which are unable to move into the capillaries, are destroyed, resulting in the hemolytic anemia associated with this disease. Individuals who are heterozygous for the HbS mutation are, for the most part, not affected with the disease. They do, however, show a greater resistance to malarial infections, and if they develop a malarial infection it tends to be a much milder form of malaria. This phenotype gives the heterozygote a selective advantage in a geographic area where malaria is a common problem and is no doubt related to the high frequency of sickle cell anemia in parts of Africa and Asia. Cystic fibrosis (CF), the most common potentially fatal autosomal recessive disorder found in Caucasians, occurs with a frequency of 1 in 2500 live births. The disease generally affects children and young adults, who suffer from mucus accumulation obstructing their airways. Infections with *Pseudomonas* and *Staphylococcus* organisms can often lead to respiratory failure. The mucus accumulation also can obstruct the pancreatic ducts as well as cause problems with the intestines and the liver. Men with the disease tend to be infertile. Patients with CF have excessive salt loss in their sweat and can be diagnosed for the disease by the "sweat test."

Lysosomal Storage Diseases: Tay-Sachs Disease and Gaucher Disease

Intracellular cytoplasmic particles known as lysosomes contain numerous enzymes that are required for the digestion of many types of macromolecules. Among these macromolecules are glycoproteins, glycosaminoglycans, and glycolipids, all of which have complex carbohydrates attached. Removal of these carbohydrate residues is catalyzed by specific sugar hydrolases (Figure **C**), which, when defective, lead to the accumulation of their respective substrates. Mutations that result in a defect in the enzymes involved in the individual degradation steps give rise to several different diseases, including Tay-Sachs disease (hexosamidase A, step 1) and Gaucher disease (β-glucosidase, step 3). Disorders of glycolipid metabolism are most common among individuals of Ashkenazi Jewish descent.

Phenylketonuria (PKU)

The basic biochemical defect in classic PKU is the loss of function of the enzyme phenylalanine hydroxylase, which catalyzes the conversion of phenylalanine to tyrosine (Figure **D**). The lack of PKU results in excess levels of phenylalanine, leading to severe mental retardation if the condition is not treated early in life.

For more information see Hoffee PA: *IMS: Medical Molecular Genetics.* Fence Creek, pp. 240–241, 260–269.

27 Autosomal Recessive Inheritance II

Common Mutations Associated with Some Disorders

A Common mutations associated with some autosomal recessive inherited disorders

Disorder	*Protein altered*	*Mutation*
Sickle cell anemia	β-Globin	Glu-to-Val missense mutation
Cystic fibrosis	CFTR	ΔF508 3-bp deletion mutation
Tay-Sachs disease	Hexosaminidase A	4-bp insert in exon 11 (a frameshift mutation) Splicing mutation in intron 12 Gly-to-Ser missense mutation in exon 7
Gaucher disease	β-Glucosidase	L444P in exon 10 (a missense mutation) IVS2+1 (a splicing mutation) 84GG (a frameshift mutation) N370S in exon 9 (a missense mutation)
Phenylketonuria	Phenylalanine hydroxylase	IVS12nt1 (a deletion of exon 12 and stop mutation) R158Q (a missense mutation) R261Q (a missense mutation) R408W (a missense mutation)

Note: CFTR = transmembrane conductance regulator; Glu = glutamine; Gly = glycine; Ser = serine; Val = valine

B Cystic fibrosis gene and gene product

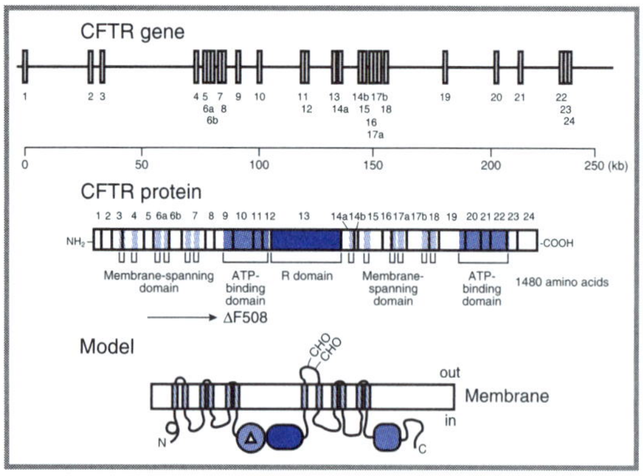

C β-Glucosidase gene

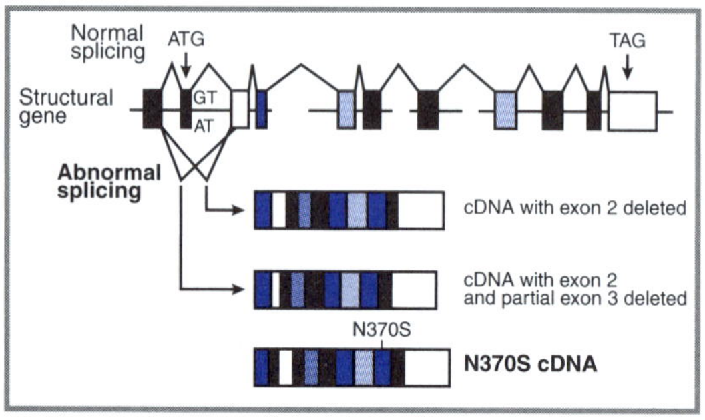

D Phenylalanine hydroxylase gene

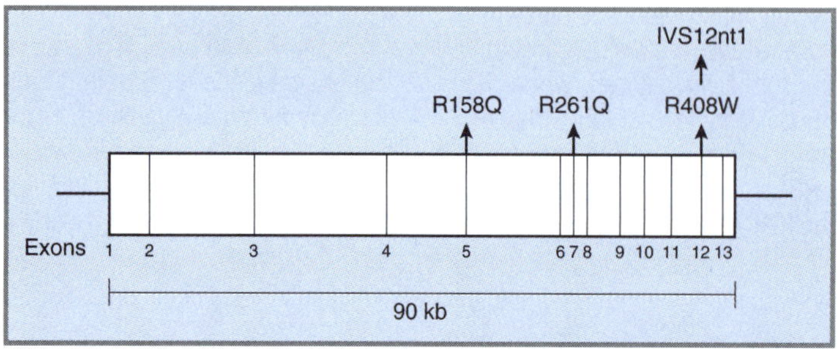

OVERVIEW

Many of the genes that are associated with autosomal recessive inherited disorders have been isolated and studied at the molecular level. For some disorders, certain common mutations occur with a high frequency among patients with a particular disease (Figure **A**). In some cases, the common mutations are attributed to a founder effect, in which a particular mutation is inherited from a common ancestor and persists in a particular population.

Cystic Fibrosis

The CF gene, isolated in 1989, is made up of 27 exons spanning 230 kb of DNA and codes for a 6.5 kb mRNA that contains information for a protein made up of 1480 amino acids (Figure **B**). This protein is known as the cystic fibrosis transmembrane conductance regulator (CFTR). Comparison of cDNA base sequences taken from normal individuals and CF patients revealed a specific 3-bp deletion in exon 10 that leads to a loss of a phenylalanine residue at position 508 in the protein (Δ F508). This particular mutation is associated with 70% of the mutant alleles in CF and is an example of a founder effect. Further studies identified more than 350 additional mutations within the gene that are associated with the remaining 30% of the mutant alleles of CF. The CFTR is glycosylated protein with a molecular mass of 170,000 and contains two hydrophobic transmembrane domains and two nucleotide-binding folds. In addition, the protein has a regulatory (R) domain containing serine residues that are the target of protein kinase A phosphorylation. The normal protein functions as a cAMP-regulated chloride channel, with the transmembrane domains spanning the epithelial cell membrane as part of the channel pore. Mutations in the CF gene can result in a complete loss of the protein, a defect in protein processing that occurs with the Δ F508 mutation, a defective regulation of the channel, or a defective conduction through the channel.

Tay-Sachs Disease and Gaucher Disease

Mutations in the hexosaminidase A (*Hex A*) gene are associated with Tay-Sachs disease. Two specific mutations in the *Hex A* gene account for the severe infantile form of Tay-Sachs disease that occurs in individuals of Ashkenazi Jewish descent (see Figure **A**). One mutation, found in 79% of carriers, is the result of a 4-bp insertion in exon 11, which leads to a frameshift mutation and a downstream stop codon. The second mutation, found in 15% of carriers, involves the donor splice junction in intron 12, resulting in the production of aberrant mRNAs that are improperly spliced. A third mutation, found in 3% of carriers, is a glycine-to-serine missense mutation in exon 7. This mutation gives a late-onset adult form of the disease in which the mutant protein is produced but is unstable, resulting in the amount of enzyme activity being reduced sufficiently to result in a clinical phenotype. In other populations, the 4-bp insertion in exon 11 is seen in 20% of non-Jewish carriers, with the exon 7 missense mutation present in 5% of non-Jewish carriers. Approximately 15% of non-Jewish carriers have a splice site mutation in intron 9 that is different from the splice site mutation common in Ashkenazi Jewish populations. The β-glucosidase gene, mapping to chromosome 1q21, codes for β-glucosidase, the enzyme defective in Gaucher disease (Figure **C**). The gene contains 11 exons spanning 32 kb of DNA and codes for 2.5 kb mRNA and a membrane-associated glycoprotein that is 497 amino acids in length with a molecular weight of 55,575–65,000, depending on its state of glycosylation. Over 30 different mutations that cause Gaucher disease have been identified in the β-glucosidase gene. These loss-of-function mutations include missense, nonsense, insertions, deletions, and splice-site mutations. Only four of these, however, occur with any frequency among individuals with Gaucher disease. A missense mutation in exon 10 (L444P), in which a leucine is changed to a proline residue at amino acid 444, is found most often in the type III Gaucher disease that occurs in the Swedish population. Among the Ashkenazi Jewish population, 70% of individuals with type I Gaucher disease have a missense mutation in exon 9 (N370S), in which an asparagine is changed to a serine residue at amino acid 370. Approximately 12% of the patients have a single base insertion (G) (84GG) to give a frameshift mutation at position 1035 that leads to a severe type II form of Gaucher disease. Approximately 4.5% have the L444P mutation, and 1%–2% have a base substitution at position 1067 (IVS2+1), which results in aberrant splicing and type II Gaucher disease (see Figure **A**). Molecular analyses indicate that the carrier frequencies for Gaucher disease alleles in the Ashkenazi Jewish population could be as high as 1 in 10–12 individuals, leading to a disease frequency of 1 in 400–600. The severity of the disease may be related to the basic mutation and whether any residual enzyme activity is present in the lysosomes.

Phenylketonuria

The phenylalanine hydroxylase gene codes for a protein with 452 amino acids and a molecular weight of 51,800 (Figure **D**). The gene, which maps to 12q22–q24 and spans 90 kb of DNA, contains 13 exons and codes for a mature mRNA of 2.4 kb. Close to 200 different loss-of-function mutations that result in PKU have been identified in the gene. The classic form, or severe form, of PKU is found in individuals with very low or undetectable levels of phenylalanine hydroxylase. Milder forms of the disease are a result of missense mutations, in which a protein is produced but has reduced activity. Several specific mutant alleles are found in European individuals with PKU and include a splicing mutation in intron 12 and three different missense mutations: R158Q, R261Q, and R408W (see Figure **A**).

For more information see Hoffee PA: *IMS: Medical Molecular Genetics*. Fence Creek, pp. 260–269.

28 Autosomal Recessive Inheritance III

Clinical Aspects of Some Disorders

A Clinical significance: Cystic fibrosis, Tay-Sachs disease, Gaucher disease, and PKU

Disease	Characteristics and Diagnostic Methods
Cystic fibrosis	Common mutation, delta F508, diagnosed with PCR and ASOH
Tay-Sachs disease	Cherry red spot in eye, increase in GM_2 ganglioside, common mutations diagnosed with PCR and ASOH
Gaucher disease	Type I: adult and most common, nonneuropathic, enlargement of liver and spleen Type II: infantile, severe form, neurologic defects, death in 2 years Type III: juvenile, progressive, intermediate severity, common mutations detected by PCR and ASOH
Phenylketonuria	Increase in phenylalanine levels, newborn screening by the Guthrie test

B Guthrie test

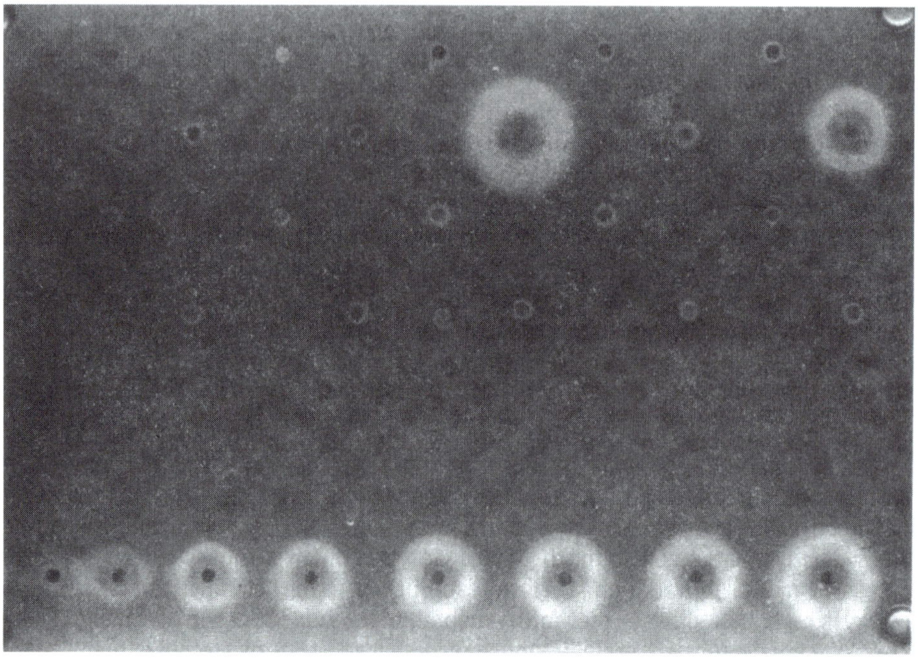

OVERVIEW

For many of the autosomal recessive inherited disorders, the type of mutation that occurs in the gene can be related to the severity of the clinical phenotype. For example, mutations such as nonsense mutations or deletion mutations that result in complete loss of the protein product often give a more severe clinical phenotype than missense mutations that may lead to a protein product with some residual enzymatic activity. Additionally, common mutations associated with many of these diseases (see Chapter 27) allow for genetic screening and testing to be done at the DNA level to determine the presence of the mutation in an individual.

Clinical Significance: Cystic Fibrosis

Diagnosis of CF is based primarily on clinical presentation and the sweat test (Figure **A**). Molecular diagnostic methods are available for detecting the Δ F508 mutation. They generally involve polymerase chain reaction (PCR) amplification followed by allele-specific oligonucleotide hybridization (ASOH). These tests are most important in genetic counseling; population screening for CF is being considered even though screening for the Δ F508 mutation alone will detect only 70% of carriers. Some 10–30 other mutations in the CF gene can be tested for by PCR and ASOH, allowing approximately 90% of the mutations associated with CF to be detected. In families with a 1 in 4 risk of having an affected child, fetuses can be tested by chorionic villus biopsy followed by DNA analysis of the DNA samples. Newborns can be tested by obtaining a cheek brushing, which is used for molecular analysis. Current treatment of CF patients involves percussion of the chest, antibiotic treatment to prevent infection, and sometimes pancreatic enzyme replacement.

Clinical Significance: Tay-Sachs Disease

Mutations that result in a complete loss of hexosaminidase A activity lead to the infantile form of Tay-Sachs disease, which is a particularly devastating form of the disease (see Figure **A**). Children born with infantile Tay-Sachs disease are generally asymptomatic before 3 months of age, at which time they begin to develop mild motor weakness, followed by a development of progressive neurological degeneration. Death occurs at 2–4 years of age. A characteristic of infantile Tay-Sachs disease is the appearance of a cherry red spot in the eye of the infant. Additionally, there is an accumulation and precipitation of GM_2 ganglioside in the brain. In the diagnosis of Tay-Sachs disease, assays for hexosaminidase A are carried out on synthetic substrates or GM_2 ganglioside, using tissue samples, serum samples, or leukocytes obtained from patients. With the identification of common mutations causing Tay-Sachs disease, molecular diagnosis can be carried out using PCR amplification and ASOH. Treatment for Tay-Sachs disease is limited to management of the patient's problems involving infections and nutrition.

Clinical Significance: Gaucher Disease

Gaucher disease, which can vary from a mild to a severe form, is classified into types I, II, and III. Type I Gaucher disease, the most common form, is nonneuronopathic and shows no primary involvement of the central nervous system (CNS) (see Figure **A**). This form of the disease is characterized by enlargement of the spleen and liver and by skeletal disease. There is a broad range of severity, from mildly affected adults to severely affected children who may die within the first 2 decades of life. The children suffer from abnormalities of the liver and spleen and have skeletal defects. Type II Gaucher disease, the most severe form of the disease, is characterized by extensive neurologic defects and leads to death of affected infants in the first 2 years of life. Type III Gaucher disease shows an intermediate severity with involvement of the visceral organs as well as some neurologic symptoms. However, the symptoms appear later than those of type II and are much less severe, with patients often surviving to young adulthood. Gaucher disease can be diagnosed by an enzymatic assay for β-glucosidase that is carried out using peripheral blood cells. The assay is not able to differentiate types II and III Gaucher disease from type I. However, DNA technology is available to detect the four common mutations, which account for nearly 95% of the alleles in Ashkenazi Jewish individuals with Gaucher disease. In non-Jewish populations, mutations L444P and N370S account for 75% of the disease alleles. For an at-risk family there is an excellent chance of detecting the mutant alleles using PCR and ASOH. Once an allele is detected, it is often possible to predict the prognosis of the disease. Homozygotes for the L444P mutation will probably have a severe visceral and neurologic disease. Individuals in whom the N370S mutant enzyme is present are generally not at risk for neurologic involvement. Bone marrow transplantation is available for treating the disease but generally is restricted to the most severe cases. Enzyme replacement therapy has been used to treat type I Gaucher disease with some success, but the procedure is very expensive. β-glucosidase can be measured in amniotic fluid and chorionic villus samples, making prenatal diagnosis a valuable tool, particularly for those families at high risk.

Clinical Significance: Phenylketonuria

Newborns are routinely screened for PKU by a bacterial assay known as the Guthrie test (Figure **B**; see Figure **A**). This test is carried out on a blood spot and measures the amount of phenylalanine in the blood. Treatment of PKU requires early intervention within a few months of birth, since seizures begin at 6–12 months of age. Infants with high levels of phenylalanine are given a low-phenylalanine diet by 4 weeks of age, and the diet is maintained sometimes for life. It is particularly important for women who have PKU to maintain a restrictive diet during pregnancy to prevent damage to the fetus.

For more information see Hoffee PA: *IMS: Medical Molecular Genetics*. Fence Creek, pp. 260–269.

29 Autosomal Dominant Inheritance I

General Concepts

A Typical autosomal dominant pedigree

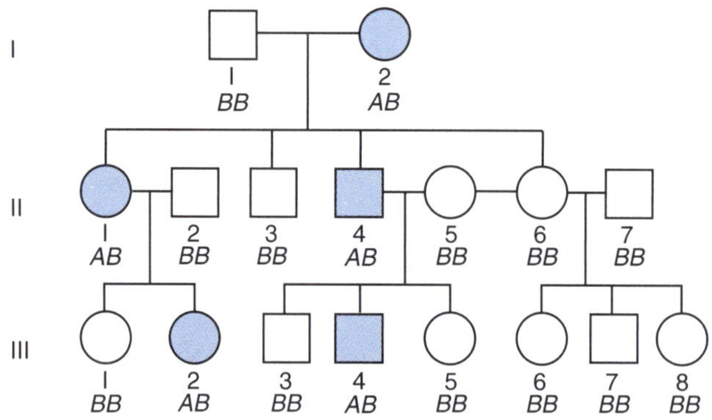

		Affected Parent (AB)	
	Gametes	*A*	*B*
Unaffected Parent (BB)	*B*	*AB*	*BB*
	B	*AB*	*BB*

Note. Shaded area signifies affected offspring genotypes.

B Autosomal dominant inherited disorders

Disorder	Protein Altered	Frequency
Change of function		
Huntington disease	Huntingtin	1/10,000
Myotonic dystrophy	Protein kinase	1/10,000
Osteogenesis imperfecta	Collagen	1/10,000
Haploinsufficiency		
Familial hypercholesterolemia	Low-density lipoprotein	1/500
Predisposition to developing cancer		
Retinoblastoma	pRb	1/15,000
Li-Fraumeni syndrome	p53	Rare
Early-onset breast, ovarian, prostate, colon cancer	BRCA1	1/200
Early-onset breast cancer	BRCA2	1/200
Colon cancer (FAP)	APC	1/7000
Colon cancer (HNPCC)	MSH-2, MLH-1,PMS-1, PMS-2, and GTBP	1/500
Neurofibromatosis 1	Neurofibromin (GTPase)	1/4000
Neurofibromatosis 2	Merlin/schwannomin	1/40,000

Note. FAP = familial adenomatous polyposis; HNPCC = hereditary nonpolyposis colon cancer; GTPase = guanosine triphosphatase.

OVERVIEW

In autosomal dominant inheritance, clinical symptoms are associated with a defect in only one allele of a gene. In these disorders, even if only one parent is affected with the disorder, the risk that a child will be affected is 50%. Change/gain-of-function mutations, caused by a mutation in the DNA that results in the production of an abnormal gene product, often lead to dominant inheritance. These mutations are much more restricted than loss-of-function mutations because they require that a protein or RNA be produced and remain stable even though it is functionally abnormal. Many of the change/gain-of-function mutations are due to a missense mutation occurring in the coding region of a gene, resulting in the production of a mutant protein that now has an altered function. In some instances, the presence of this mutant protein can result in a dominant-negative phenotype, in which the abnormal protein inactivates or inhibits the gene product of the normal allele. Dominant-negative effects usually involve proteins that form oligomers, in which case the mutant subunit forms a complex with the normal subunits and the resulting oligomer is inactive or altered in its function. Another type of change/gain-of-function mutation that shows dominant inheritance is represented by the trinucleotide expansion disorders, in which expansions of the trinucleotide repeats within a gene can result in a protein or RNA being produced, but one that functions in an abnormal way.

Autosomal Dominant Pedigrees

A typical autosomal dominant pedigree is shown in Figure **A** and illustrates the four main characteristics of diseases that require the presence of only a single defective allele to produce an affected phenotype. (1) There is a 50% chance that any child of an affected individual will be affected with the disease. If the disease has two alleles, *A* and *B*, with *A* as the affected dominant allele, then both *AA* and *AB* offspring will be affected. Generally, because the majority of affected individuals are heterozygotes, most matings will be between an unaffected individual, *BB*, and an affected heterozygote, *AB*. In this case, half of the children will be *AB* and affected, and half will be *BB* and unaffected. (2) Because only a single allele is required for an individual to have the affected phenotype, affected individuals are seen in every generation. (3) Unaffected individuals do not carry the defective allele and thus do not transmit the affected phenotype to their offspring. (4) Males and females have an equal chance of receiving the defective allele and being affected.

Autosomal Dominant Inherited Disorders

Dominantly inherited disorders fall into several main classes based on the type of mutation that has occurred and the effect that the mutation has on the phenotype (Figure **B**). They include change-of-function mutations, in which the product of the gene has an altered function; gain-of-function mutations, in which the product of the gene has acquired a new function; dominant-negative mutations, in which the product of the gene interferes with the function of the protein produced by the normal allele; and haploinsufficiency, in which the amount of gene product produced by the normal allele is insufficient to prevent a clinical phenotype. Additional diseases that show autosomal dominant inheritance are those that result in a predisposition to develop cancer. Most of these diseases result from mutations in tumor suppressor genes (see Chapter 22). Even though the defective allele of the tumor suppressor gene is recessive to the normal allele, the inheritance of a single defective allele now predisposes the individual to cancer if a second mutation occurs at the somatic cell level.

Change-/Gain-of-Function and Dominant-Negative Inherited Diseases.

In human diseases that result from the expansion of trinucleotide repeats, repeats of trinucleotides that occur at the beginning or end of a gene undergo expansion and alter the protein or the RNA coded for by the gene. This expansion can occur in the 5′ or 3′ noncoding regions of the gene or within the first exon or first intron of the gene. The site of the expansion determines the effect of the mutation on the RNA or protein product. Two diseases, Huntington disease and myotonic dystrophy, are both a result of trinucleotide expansion and are both inherited as an autosomal dominant trait. Osteogenesis imperfecta, a disorder of collagen biosynthesis, is an example of an autosomal dominant inherited disease that has a dominant negative effect on the protein produced by the normal allele.

Haploinsufficiency.

Genes localized to autosomal chromosomes are present in two copies in humans. If one of the alleles is rendered defective (i.e., the heterozygous state), the individual generally is unaffected with the particular disease associated with that defective allele. In these cases, the heterozygote is a carrier and is usually clinically asymptomatic, even with only 50% of the normal gene product. Exceptions to this are the autosomal dominant disorders, such as myotonic dystrophy and type II osteogenesis imperfecta, in which the abnormal gene product interferes with the function of the normal gene product. In these cases, the mutant phenotype is dominant to the normal phenotype, resulting in a clinical disease in the heterozygous state. Another exception occurs when 50% of the normal gene product is not enough for a normal phenotype, as is found in type I osteogenesis imperfecta. Although type I osteogenesis imperfecta is generally a mild form of the disease, there is an obvious clinical phenotype. Diseases in which 50% of the normal amount of protein is not enough to prevent a clinical phenotype in the heterozygous state are called haploinsufficiency diseases. One of the most important diseases that shows haploinsufficiency and is inherited as an autosomal dominant trait is familial hypercholesterolemia.

For more information see Hoffee PA: *IMS: Medical Molecular Genetics*. Fence Creek, pp. 239–240, 270–282.

30 Autosomal Dominant Inheritance II

Trinucleotide Expansion Disorders

A Autosomal dominant diseases with trinucleotide expansions

Disease	Protein	Repeats: Normal	Repeats: Mutant	Repeats: Type
Huntington disease[a]	Huntingtin	11–34	36–120	Glutamine (CAG) repeats[a]
DRPLA[a]	Atrophin	7–23	49–75	"
Spinocerebellar ataxia (type I)[a]	Ataxin-1	6–44	40–82	"
Spinocerebellar ataxia (type 3)[a]	MJD-1	13–40	68–69	"
Myotonic dystrophy[b]	Protein kinase	5–30	50–2000	CTG repeats

[a] Glutamine repeats (CAG) in the coding region of the first exon of the gene.
[b] CTG repeats in the noncoding 3′ end of the last exon of the gene.
Note. Dentatorubal-pallidoluysian atrophy. MJD = protein defective in spinocerebellar ataxia type 3, which is also known as Machado-Joseph disease.

B Position of the trinucleotide repeats in Huntington disease and myotonic dystrophy

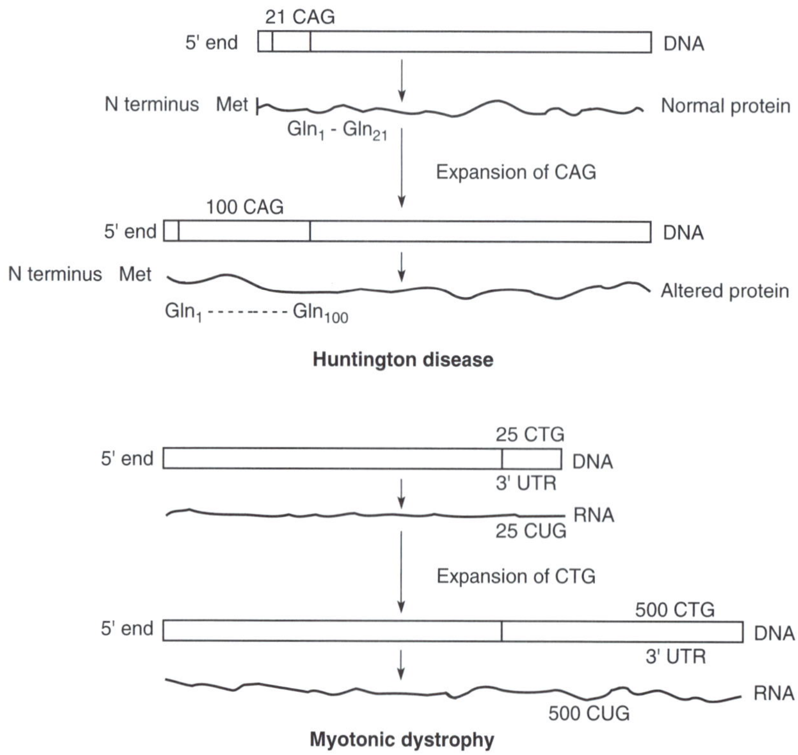

OVERVIEW

The trinucleotide, CAG, codes for the amino acid glutamine. In four autosomal dominantly inherited human diseases—Huntington disease, dentatorubral-pallidoluysian atrophy (DRPLA), and spinocerebellar ataxia, type 1 and type 3—a CAG repeat is located in the coding region of the first exon at the 5′ end of the gene (Figure **A**). These repeats are translated and appear in the protein as a stretch of glutamines. A normal protein will have a repeat of glutamines of less than 45 at its amino terminal end, whereas the mutant proteins will have a range of 36–120 glutamines at that site. All of the CAG repeat diseases involve late-onset neuronal loss and are inherited in an autosomal dominant fashion. The various proteins involved in the above diseases have no homology to one another and probably serve different functions in the cell, suggesting that it is the excess glutamine in the proteins that gives them their mutant phenotype. Other trinucleotide expansions occur in noncoding regions of genes and are not translated into protein. In the case of myotonic dystrophy, repeats of CTG, occurring in the 3′UTR of the gene coding for a protein kinase, are not translated into protein.

Huntington Disease

Huntington disease, a late-onset, progressive, fatal disease, has the trinucleotide, CAG, expanded in the coding region of the first exon of the gene, resulting in an altered huntingtin protein with multiple glutamines at its amino-terminal end. The function of huntingtin remains unknown, although the protein, which contains 3144 amino acids, appears to be a DNA-binding protein. In the first exon, at the 5′ end of the normal gene, there is a sequence of 21 copies of CAG that results in 21 glutamines being present in the N-terminal portion of the normal protein (Figure **B**). In the disease gene, the stretch of CAG repeats is expanded to 36–120 copies, resulting in the production of an altered protein that now contains 36–120 glutamine residues. The degree of expansion of the CAG repeat and the number of glutamines in the protein correlate with the severity of the disease. The biochemical defect in Huntington disease is still unknown; consequently, there is no cure or treatment for the disease. Recent studies on the potential effects that extra glutamines at the N-terminal end of proteins can have in a cell suggest that the presence of the polyglutamines results in aggregation of the proteins into clumps. These clumps precipitate in the nuclei, causing brain dysfunction. The buildup of the clumps may be gradual, accounting for the slow progression and late onset of the disease.

Myotonic Dystrophy

Myotonic (repetitive muscle firing) dystrophy (weakness and wasting) is an autosomally dominant inherited disease associated with CTG repeats in the noncoding 3′ end of the last exon of the gene coding for a cAMP-dependent kinase (i.e., dystrophia myotonica [DM] kinase). Normal individuals have a range of 5–30 CTG repeats. Individuals who show minimal effects of the disease have repeats in the range of 50–80 copies, which is increased in adult patients to a range of 100–500 copies (see Figure **A**). Early onset of the disease is associated with an expansion of the CTG repeats in a range of 500–2000 copies. The neonatal form of myotonic dystrophy shows maternal transmission, whereby the length of the expansion increases from mother to child. In the other diseases resulting from trinucleotide expansion that are inherited as autosomal dominant disorders (e.g., Huntington disease), the expanded repeats are located at the 5′ end of the gene and affect the structure of the protein product (see Figure **B**). Therefore, the inheritance pattern is consistent with a change-of-function mutation in which the mutant protein has an altered structure leading to an altered function in the cell. In the case of myotonic dystrophy, in which the expansion of the trinucleotide repeat occurs in the 3′-UTR of the gene, the repeat sequence is not translated into protein and does not lead to the production of a mutant protein (see Figure **B**). Recent results suggest that the RNA molecule with an expanded CUG repeat at the 3′ end may have a defect in polyadenylation. Furthermore, an RNA-binding protein that is specific for CUG repeats appears to accumulate in the nucleus of myotonic dystrophy patients, suggesting that poorly polyadenylated DM kinase mRNA precipitates in the nucleus and sequesters CUG-binding proteins. This action disrupts the normal processing of RNA transcripts present in the nucleus, affecting normal DM kinase mRNA function and possibly other mRNAs, resulting in the clinical variability seen in myotonic dystrophy.

Clinical Significance

Characteristics of Huntington disease include a degeneration of the basal ganglia leading to uncontrollable muscle movements, the loss of cognitive functions, and the development of psychiatric problems. The disease can be divided into four main stages that span the age of onset of approximately 36 years to clear expression of the phenotype by age 65 years. In the initial stage of the disease, the patient has small losses of muscular coordination, forgetfulness, and some personality changes. During the second stage of the disease, uncontrollable movements begin, accompanied by speech disorders. By the third stage of the disease, the patient loses control of bodily functions and is no longer able to care for him- or herself. The fourth and final stage finds the patient in a vegetative state, and finally death occurs. Because of the late onset, generally past the reproductive stage, the transmission of the disease to offspring often takes place before the disease has been diagnosed.

The symptoms of myotonic dystrophy are quite variable and are related to the age at which the patient expresses the phenotype. When expressed in newborns, the disease is characterized by respiratory problems that result in death if the infant is not given ventilation assistance. In childhood, the expressed phenotype is characterized by mental retardation. Adults may demonstrate some myotonia and muscle weakness, and some older adults may develop only cataracts. The usual presentation of the disease is a development of progressive muscular dystrophy that is often induced by cold and associated with an inability to relax the muscles. Patients are advised to avoid anesthesia during surgery because many of them have a high sensitivity to anesthetics and may undergo heart block and sudden death.

For more information see Hoffee PA: *IMS: Medical Molecular Genetics.* Fence Creek, pp. 270–272.

31 Autosomal Dominant Inheritance III

Osteogenesis Imperfecta — a Disorder of Collagen Biosynthesis

A Structure of type I collagen

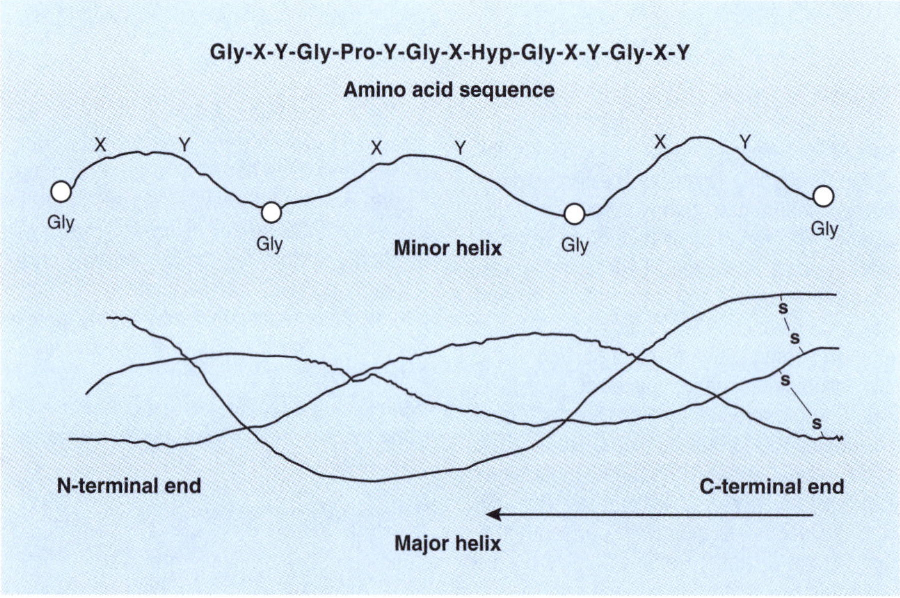

B Osteogenesis imperfecta: effect of mutations on collagen formation

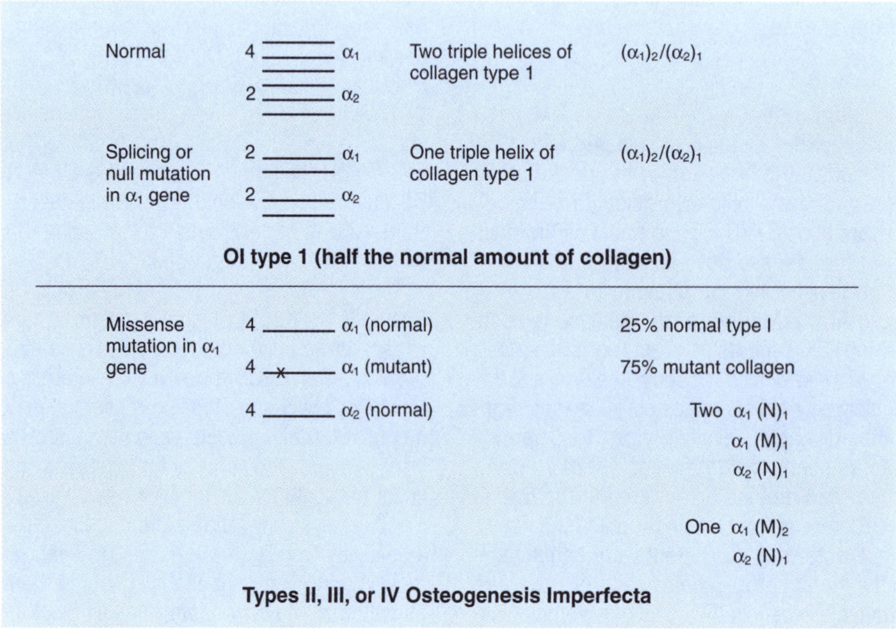

OVERVIEW

Collagen is coded for by at least 28 different genes and is found in the body in many forms, designated from type I to type XIV. Type I collagen, which is found in skin, bone, tendons, and ligaments, is altered in individuals with the autosomal dominant disorder osteogenesis imperfecta (OI).

Structure of Type I Collagen

Type I collagen is composed of three α chains (two α_1 chains and one α_2 chain). Each α chain forms a left-handed helix that is made up of three amino acids per turn (Figure **A**). Three of these left-handed helices come together to form a right-handed triple helical structure. There are three amino acids per turn in the left-handed helix, and every third amino acid is glycine, giving a repeat of Gly-X-Y, where X and Y represent other amino acid residues. Approximately 10% of the X residues are proline, and approximately 10% of the Y residues are hydroxyproline, accounting for the rigidity of the collagen molecule. Analysis of the sequence of the exons that code for collagen show that the sequence coding for Gly-X-Y is repeated numerous times and found in all of the exons. The small size of the glycine residue in the structure allows the amino acid to fit into the central core of the helix. This is extremely important, as demonstrated by the finding that many missense mutations that result in a substitution of the glycine residue by another amino acid result in the production of a mutant collagen. The final triple helical molecule, made up of two α_1 chains and one α_2 chain, is held together at the C-terminal end of the protein by interchain disulfide bonds. The formation of the triple helix begins at the C-terminal end and moves toward the amino-terminal end of the molecule. This directionality of the helix formation appears to be directly related to the effect that different missense mutations have on the severity of clinical phenotype of OI.

Osteogenesis Imperfecta

In a normal individual, two α_1 helixes and one α_2 helix come together to form the right-handed triple helix, or type I collagen. In an individual with a splicing or null mutation in one of the α_1 chain alleles, a normal molecule of type I collagen is made, but at one-half the normal amount because of the reduced level of α_1 chain (Figure **B**). Individuals with splicing or null mutations are affected with the mild type I form of OI, characterized by blue sclerae and brittle bones that can lead to as many as 50 fractures before puberty. Hearing loss is present in about one-half of the patients, but there is no skeletal deformity. The frequency of type I OI is 1 in 10,000. The molecular basis for the decrease of α_1 protein can vary from one patient to another, but one common form of the disease is known to be due to a 5-bp deletion near the 3′ end of the gene, COL1A1. This 5-bp deletion results in a frameshift mutation that extends the length of the protein a predicted additional 84 amino acids. The increased size of the protein appears to make it unstable, accounting for the deceased level of type I collagen. In addition, any mutation that results in improper splicing often gives rise to type I disease. In the other three types of OI (types II, III, and IV), the molecular defects are usually missense mutations that result in a substitution of a crucial glycine residue with another amino acid, leading to a disruption of the helix. The severity of OI is a reflection of where the substitution occurs and which amino acid is placed in the mutant α_1 chain.

Type II OI, a lethal form of the disease in infants, presents with severe deformity and death within 1 month. In type II, the missense mutations that result in a substitution of a glycine residue in the triple-helical domain of the protein generally occur in the C-terminal end of the gene and result in an altered protein that interferes with normal assembly. The normal allele is still producing a normal α_1 molecule; therefore, 25% of the triple helical molecules formed will be normal (see Figure **B**). However, since the composition of the triple helix is made up of two α_1 chains, the presence of the abnormal molecule will result in the formation of three abnormal triple helices for each normal molecule. Infants with the lethal form, type II OI, appear to have new mutations rather than an inherited mutation from one of the parents. The exception to this is when parental mosaicism occurs. Death of type II OI patients is usually a result of pulmonary complications. The disease can be diagnosed during 14 and 18 weeks of gestation by ultrasound screening, which reveals abnormal limb length and morphology. Abnormal collagen can also be detected by analysis of the chorionic villi.

Type III OI is characterized by a progressive deformation of the bone, limited growth, and hearing loss.

Type IV OI causes moderate deformity, fractures, and hearing loss. The decreased severity of type III and type IV OI, compared with type II OI, appears to be related to which glycine in the α_1 chain is changed and which other amino acid is substituted (see Figure **A**). In general, the closer the substitution is to the amino-terminal end of the molecule, the less severe the clinical phenotype, although the type of substitution is also important in determining the clinical severity. Mutations in the α_2 chain also arise, but they tend to cause less severe forms of the disease than mutations in the α_1 chain. This is probably a reflection of the presence of only one α_2 chain per triple helix, so that these individuals are still capable of producing 50% normal molecules.

For more information see Hoffee PA: *IMS: Medical Molecular Genetics*. Fence Creek, pp. 272–274.

32 Autosomal Dominant Inheritance IV

Familial Hypercholesterolemia and Haploinsufficiency

A Familial hypercholesterolemia-A defect in the low-density lipoprotein (LDL) receptor

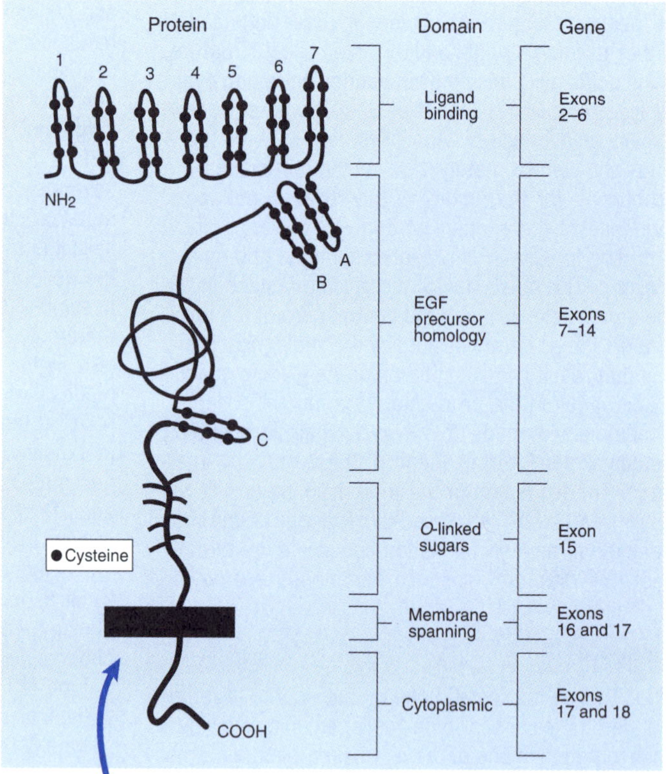

B Synthesis and function of the LDL receptor

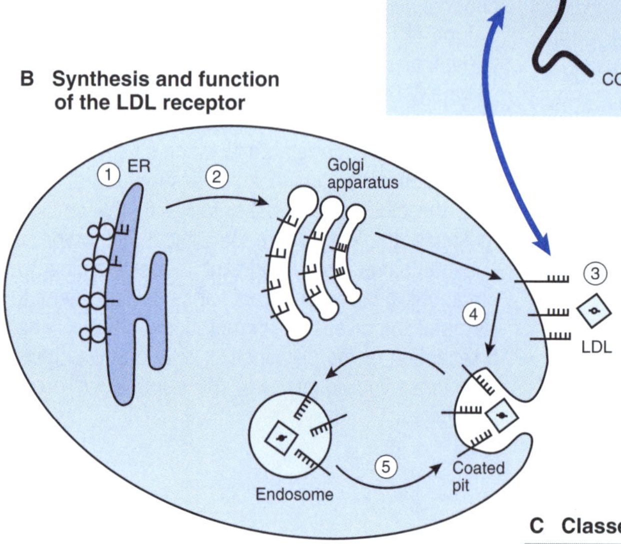

C Classes of LDL receptor mutants

Mutation class	Synthesis	Transport	Binding	Clustering	Recycling
1	→ X				
2	———	→ X			
3	———	———	→ X		
4	———	———	———	→ X	
5	———	———	———	———	→ X

OVERVIEW

Generally, humans heterozygous for a defective allele on an autosomal chromosome do not express the clinical phenotype associated with the gene defect. Exceptions to this are the autosomal dominant disorders in which the abnormal gene product interferes with the function of the normal gene product (e.g., myotonic dystrophy, type II osteogenesis imperfecta) and in diseases in which 50% of the normal gene product is not sufficient to preclude clinical symptoms. Diseases that show clinical symptoms in the heterozygous state as a result of an insufficient level of a gene product are called haploinsufficiency diseases (i.e., half is not enough). An important haploinsufficiency disease that is inherited as an autosomal dominant trait is familial hypercholesterolemia.

Familial Hypercholesterolemia—A Defect in the Low-Density Lipoprotein (LDL) Receptor

Familial hypercholesterolemia in the heterozygous state has an incidence of 1 in 500 individuals and accounts for 5% of the coronary artery disease in the Western world. Individuals with familial hypercholesterolemia have myocardial infarctions before reaching the age of 55 years and have cholesterol levels twice that of a normal individual. The disease in the homozygous state has a frequency of 1 in a million. Homozygous individuals suffer myocardial infarctions before the age of 20 years and have cholesterol levels 3–4 times the normal level. Cholesterol is carried by LDL and is taken into cells by the attachment of the LDL-cholesterol complex to the LDL receptors located in the coated pits on the surface of cells. In patients with familial hypercholesterolemia, the number of normal LDL receptors is reduced, resulting in an increase in the levels of cholesterol in the plasma. More than 300 mutations in the LDL receptor gene have been identified as typical loss-of-function mutations (i.e., missense, nonsense, insertions, deletions). The position of the mutation within the gene alters the receptor protein differently, depending on which of the LDL receptor protein functions is impaired. The LDL receptor gene maps to chromosome 19p13 and is made up of 18 exons spanning 45 kb of genomic DNA. The mature mRNA is 5.3 kb in length and codes for a protein of 839 amino acids. As indicated in Figure **A**, various exons within the gene code for different domains in the protein.

Synthesis and Function of the LDL Receptor

The LDL receptor is synthesized in the endoplasmic reticulum (ER) and then transported first to the Golgi and then to the cell surface (Figure **B**). Exons 7–14 (see Figure **A**) contain the information for the movement of the LDL receptor from the ER; exons 16–17, information for membrane spanning of the protein from inside to outside the cell; and exons 17–18, information for its localization to the coated pits on the cell surface. Once on the cell surface, the LDL receptor binds the protein moiety of the LDL-cholesterol complex, and this complex is brought into the cell by endosome invagination. The endosomes carry the LDL-cholesterol complex to the liposomes with the release and recycling of the LDL receptor to the surface of the cell, where it now binds more LDL-cholesterol complexes. Exons 2–6 contain the information for the binding of the receptor to LDL. Exon 1 is required for the signal sequence necessary for processing of the receptor protein, and exon 15 contains the sequences necessary for post-translational modification by glycosylation.

Classes of LDL Receptor Mutants

Mutations that occur in the specific exons of the LDL gene alter specific functions of the LDL receptor protein. These mutations result in familial hypercholesterolemia and have been grouped into five classes, each class representing a defect in different stages involved in the synthesis or function of the LDL receptor protein (Figure **C**). Class 1 mutations are null mutations (nonsense, frameshift, or splicing mutations) in which no protein is made by the mutant allele; an individual heterozygous for one of these mutations has 50% of normal LDL receptors. Class 2 mutations occur in exons 7–14. Individuals with one of these mutations synthesize normal amounts of a defective LDL receptor that stays in the ER and is degraded there. Mutations of Class 3 occur in exons 2–6. Individuals with one of these mutations make a defective LDL receptor that no longer can bind LDL. Class 4 mutations occur in exon 18. Individuals with one of these mutations produce a LDL receptor protein with a defective cytoplasmic tail that interferes with the ability of the receptor to internalize the bound LDL. Class 5 mutations interfere with the dissociation of the receptor from the LDL and block the recycling of the receptor back to the surface of the cell. The basic effects of all of these mutations is a reduction by one-half of the amount of the normal LDL receptors. Half is not enough, and heterozygous individuals are affected with familial hypercholesterolemia.

Clinical Significance

The elucidation of the function of the LDL receptor and the identification of the mutations and their effects on this protein have increased our understanding of this form of coronary heart disease. Many patients now can be identified by a family history of early heart attacks and elevated cholesterol levels. Treatment involves a change in diet to lower the level of cholesterol as well as medications that interfere with the synthesis of cholesterol in the liver.

For more information see Hoffee PA: *IMS: Medical Molecular Genetics*. Fence Creek, pp. 275–277.

33 Autosomal Dominant Inheritance V

Genetic Predisposition for Developing Cancer—Retinoblastoma, Li-Fraumeni Syndrome, and Familial Adenomatous Polyposis

A Inherited nature of cancer

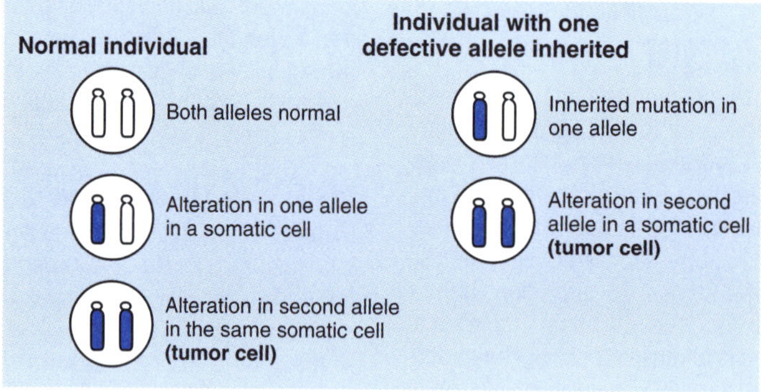

B Pedigree of a family with Li-Fraumeni syndrome

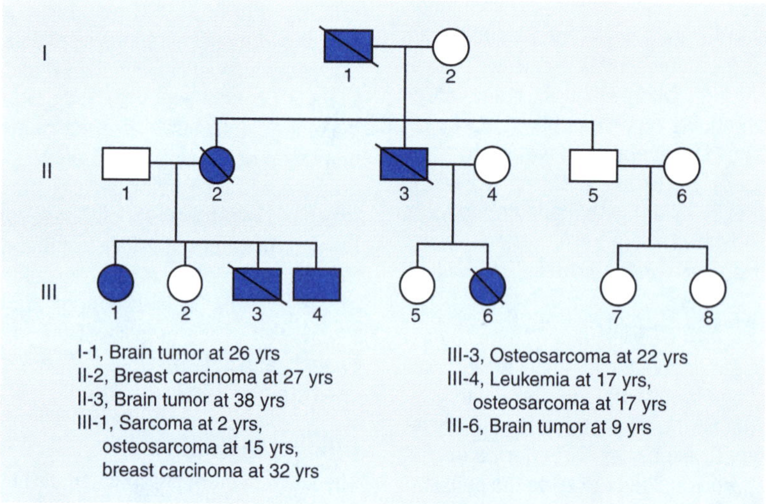

C Role of APC in growth control

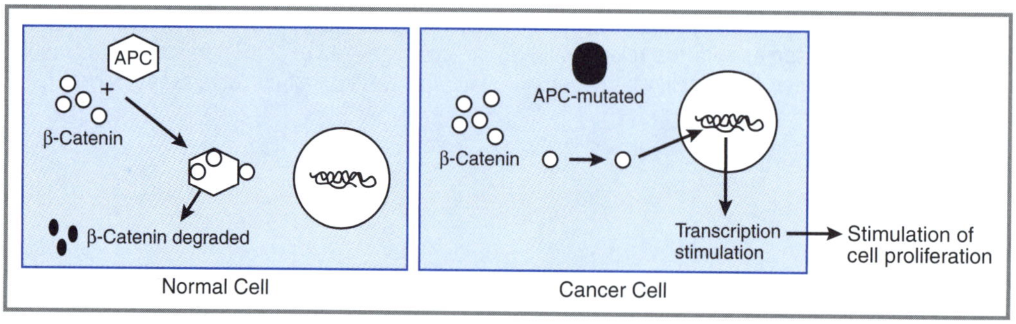

OVERVIEW

Considerable evidence has accumulated over the past 10 years indicating that families can be affected with rare autosomal dominantly inherited genetic predispositions to develop cancer. Several common cancers have a predisposition to be inherited. They include retinoblastoma (RB); Li-Fraumeni syndrome; familial adenomatous polyposis (FAP), hereditary nonpolyposis colorectal cancer (HNPCC), neurofibromatosis 1 and 2 (NF-1, NF-2), and early-onset hereditary breast and ovarian cancer (BRCA). All of these disorders, with the exception of HNPCC, are caused by mutations that lead to a loss of function of one allele of a gene that codes for a tumor suppressor protein (see Chapter 22). The cancer that develops results from a somatic cell mutation that inactivates the second allele of the tumor suppressor gene, making the cell more likely to lose growth control and develop into a malignant cell. HNPCC, discussed in Chapter 34, is not caused directly by the loss of a tumor suppressor protein but is a result of mutations that occur in genes required for DNA repair (see Chapter 14). In this case, the loss of normal DNA repair functions lead to an overall increase in mutation rate and an increased probability that a mutation will inactivate a tumor suppressor gene.

Inherited Nature of Cancer

The identification of the crucial role that tumor suppressor proteins play in the regulation of cell growth came from studies on RB, an inherited disorder expressed in children as a development of tumors of the retina. Studies on the inheritance patterns of RB by Knudson lead to the two-hit hypothesis to explain the inherited nature of RB (Figure **A**). According to this hypothesis, a single loss-of-function mutation occurs in the germ line and inactivates one allele of a crucial gene. This mutation is passed on to the offspring. A second mutation occurs in a somatic cell of the offspring, inactivating the other allele to give a homozygous state and a predisposition of the cell to become cancerous. This initial loss-of-function mutation is inherited as an autosomal recessive trait and does not in itself lead to the development of cancer. However, cells in which this single allele is inactivated will have a greater probability of developing cancer if another mutation inactivates the second allele to give a homozygous state. In a normal cell with two functional alleles, two somatic cell mutations occurring in each of the alleles are needed to produce the homozygous state and place that cell at the same risk for developing into a cancer cell. The chance that two mutations will occur in a single cell is much lower than the chance that a single mutation will occur in the cell. In some cancers, more than two mutations may be required for cancer development, but in all cases, members of a family with an initial mutation in one of the tumor suppressor gene alleles will have a greater risk of developing cancers than those without the inherited mutation.

Examples of Familial Inherited Cancers

Retinoblastoma.

The gene responsible for RB is mapped to 13q14 and codes for the protein pRb, which has a molecular weight of 110,000. pRb plays a crucial role both in cell cycle control and transcriptional regulation (see Chapter 23). Inactivation or modification of pRb will potentially disrupt the normal cell cycle controls, leading to uncontrolled growth and a potentially cancerous cell.

Li-Fraumeni Syndrome.

A study of children with rhabdomyosarcoma carried out in 1969 by Li and Fraumeni identified four families in which there was an increased incidence of soft tissue tumors, breast cancer, brain tumors, osteosarcomas, and other neoplasms. This familial cancer syndrome, known as the Li-Fraumeni syndrome, is the result of a germline mutation in the gene coding for the tumor suppressor protein, p53. A pedigree for a family with Li-Fraumeni syndrome is shown in Figure **B**. In this family, members at risk (i.e., who carry the mutation) have a 50% probability of developing cancer by 30 years of age, compared with 1% of the general population. Although the Li-Fraumeni syndrome and the inheritance of a defective p53 gene is quite rare, the presence of a mutated p53 in human cancers is quite common. More than 50% of all human cancers have a mutation in the gene coding for p53, which plays a central role in normal growth control. Loss or mutation of p53 can have multiple effects on growth regulation because p53 functions both as a transcription factor and as a controlling factor to sense DNA damage in the genome. When DNA damage occurs, p53 signals the cell to stop in the G1 phase of the cell cycle to allow DNA repair to occur (see Chapter 16). In the absence of a normal p53, a cell with damaged DNA will continue to grow and transfer the damage to daughter cells, allowing the defect to be perpetuated in subsequent generations, eventually leading to a cancerous cell.

Familial Adenomatous Polyposis.

FAP is characterized by the development of multiple benign colorectal tumors, some of which undergo additional mutations and eventually lead to the development of carcinomas. In the germline, the adenomatous polyposis coli (APC) gene located on chromosome 5q21 undergoes a mutation in one allele, which is passed on and leads to FAP in the offspring. The normal APC protein of 310,000 molecular weight interacts with, and regulates the level of, β-catenin, a protein that sends growth signals to the nucleus (Figure **C**). In the absence of normal APC protein, down-regulation of the β-catenin transcription function is lost, leading to increased cell growth and an increase in the rate of additional mutations, eventually resulting in the alteration of K-Ras and p53 functions. As in other disorders that result from mutations in tumor suppressor genes, only the initial mutation is inherited; the development of cancerous cells arises from the additional mutations that occur in somatic cells.

For more information see Hoffee PA: *IMS: Medical Molecular Genetics*. Fence Creek, pp. 276–282.

34 Autosomal Dominant Inheritance VI

Genetic Predisposition for Developing Cancer—Hereditary Nonpolyposis Colorectal Cancer, Neurofibromatosis 1 and 2

A Function of genes associated with HNPCC

Gene	Function of Protein	% of HNPCC patients defective
MSH-2	Binds mispair in DNA	60%
MLH-1	Involved in excision	30%
PMS-1	Homologue to MLH-1	
PMS-2	Homologue to MLH-1	10%
MSH-3	Forms complex with MSH-2	
MSH-6	Forms complex with MSH-2	

B Function of neurofibromin

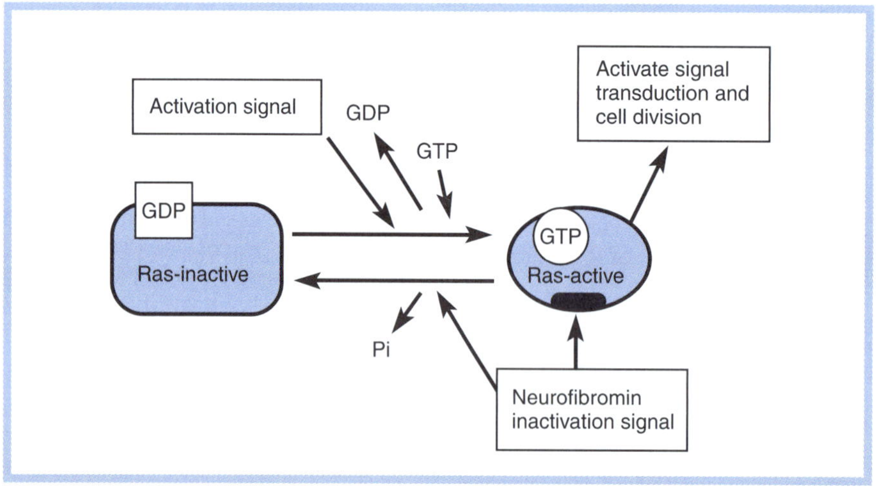

C Diagnostic criteria for neurofibromatosis 1 and 2

Disease	Protein	Diagnostic Criteria
NF-1	Neurofibromin	Six or more café-au-lait spots, larger than 15 mm in size Two or more Lisch nodules (small bumps on the iris of the eye) Two or more neurofibromas (benign skin tumors) Optic glioma Freckles within skin folds Skeletal deformity such as bowing of lower legs
NF-2	Schwannomin	Clinical symptoms appear as teenagers or older Bilateral acoustic neuromas Family history of NF-2 and either a neurofibroma, a schwannoma, a meningioma or a glioma

OVERVIEW

A second form of hereditary colon cancer is hereditary nonpolyposis colorectal cancer (HNPCC), which accounts for nearly 10% to 15% of all colon cancers. The development of HNPCC is a result of mutations in the genes involved in the methyl-directed mismatch repair (MMR) system described in Chapter 14. The MMR system removes improperly paired bases from the DNA, thus preventing mutations from occurring at the next replication cycle.

Function of Genes Associated with HNPCC

Six human genes that function to keep mismatches from persisting in the DNA have been identified (Figure **A**). These genes include: *MSH-2, MLH-1, PMS-1* and *PMS-2,* and *MSH-3* and *MSH-6.* Although mutations in each of these genes result in an inherited predisposition to develop cancer, the molecular basis for cancer development differs from that seen with the classical tumor suppressor mutations. In HNPCC, the cells have a defective MMR system, which leads to a mutator phenotype (i.e., a cell with a dramatic increase of 100–1000-fold in the overall mutation rate). The growth properties of HNPCC cells are not initially altered, but the increased mutation rate increases the likelihood that mutations in other crucial genes, such as tumor suppressor genes, will occur that will lead to the loss of growth control and cancer. One hallmark of HNPCC is microsatellite instability, in which the tri- and tetranucleotide repeat sequences vary in number. This microsatellite instability appears to be due to mistakes made by DNA polymerase when replicating these repeat sequences. Normal cells with a functional MMR system will correct the mistakes; however, cells with a defect in the MMR system will be unable to correct the defect, and the mistake will become a permanent part of the DNA sequence. As seen in some other types of inherited cancers, several mutations appear to be found in patients more often than others, indicating a founder effect. In HNPCC, a deletion of exon 5 of the *MSH-2* gene, a deletion of exon 16 of the *MLH-1* gene, and a deletion of exon 6 of the *MLH-2* gene are most common.

Neurofibromatosis

NF-1 affects approximately 1 in 4000 individuals and is characterized by the appearance of café-au-lait spots (at least six spots larger than 15 mm) and at least two Lisch nodules on the iris. Additionally, benign neurofibromas, auxiliary freckling, and optic gliomas develop. The disease, initially known as von Recklinghausen disease, can vary in its severity, but generally no significant decrease in life expectancy occurs. The disease shows a high mutation rate, and many cases (30%–50%) are caused by new mutations. Individuals with NF-1 are predisposed to some cancers at a low risk (twice the overall risk), including neurofibrosarcomas, malignant schwannomas, and gliomas. Linkage analysis mapped the NF-1 gene to 17q, and with the finding of two NF-1 patients with translocations at 17q11.2, the map position of the gene was finally pinpointed. The NF-1 gene, approximately 400 kb in size, was cloned by positional cloning and shown to code for a 2800 amino acid–containing protein known as neurofibromin (Figure **B**). The protein shows the presence of a GTPase-activating protein (GAP) domain that is homologous to two Ras inhibitor proteins found in yeast. Ras is a protein that binds GTP and is involved in the signal transduction pathway. The function of neurofibromin is to interact with GTP-bound Ras, thereby participating in the hydrolysis of GTP to GDP with subsequent inactivation of Ras. If Ras has undergone mutation to its oncogene form, it is no longer subject to inactivation by neurofibromin and can result in an abnormal or cancerous cell. When mutations occur in the NF-1 gene, the defective neurofibromin can no longer inactivate Ras, resulting in increased levels of GTP-bound Ras and an alteration in signals that control cell growth and division.

NF-2 is also inherited as an autosomal dominant disorder but at a much lower frequency (1 in 40,000) than NF-1. The disease is a result of mutations in the NF-2 gene located on chromosome 22q12 and is a separate disease from NF-1. The primary difference is the development of bilateral vestibular schwannomas (acoustic neuromas), which occur in adolescents or young adults and lead to the development of deafness and balance disorders. In addition, individuals with NF-2 tend to develop tumors of the brain, the spinal cord, and the peripheral nerves. The gene was cloned by positional cloning and shown to code for a protein that has homology to a family of cytoskeleton-associated proteins. The protein (named merlin) is also referred to as schwannomin for its involvement in schwannomas. The function of this protein is to serve as a link between the cytoskeleton and the cell membrane.

Diagnosis and Treatment of Neurofibromatosis

At present, there is no known cure for NF. The correct diagnoses of NF-1 and NF-2 are based on the criteria listed in Figure **C**. Once diagnosed, treatment may involve surgical intervention for scoliosis, plastic surgery to remove some skin tumors, or neurosurgery to treat brain or spinal tumors. Genetic counseling is important for families with NF to explain the risk to future offspring and to educate parents who have a child diagnosed with NF about potential symptoms, both psychological and physical.

For more information see Hoffee PA: *IMS: Medical Molecular Genetics.* Fence Creek, pp. 281–282.

35 Autosomal Dominant Inheritance VII

Genetic Predisposition for Developing Cancer—Hereditary Breast and Ovarian Cancer

A *BRCA1* and *BRCA2* genes

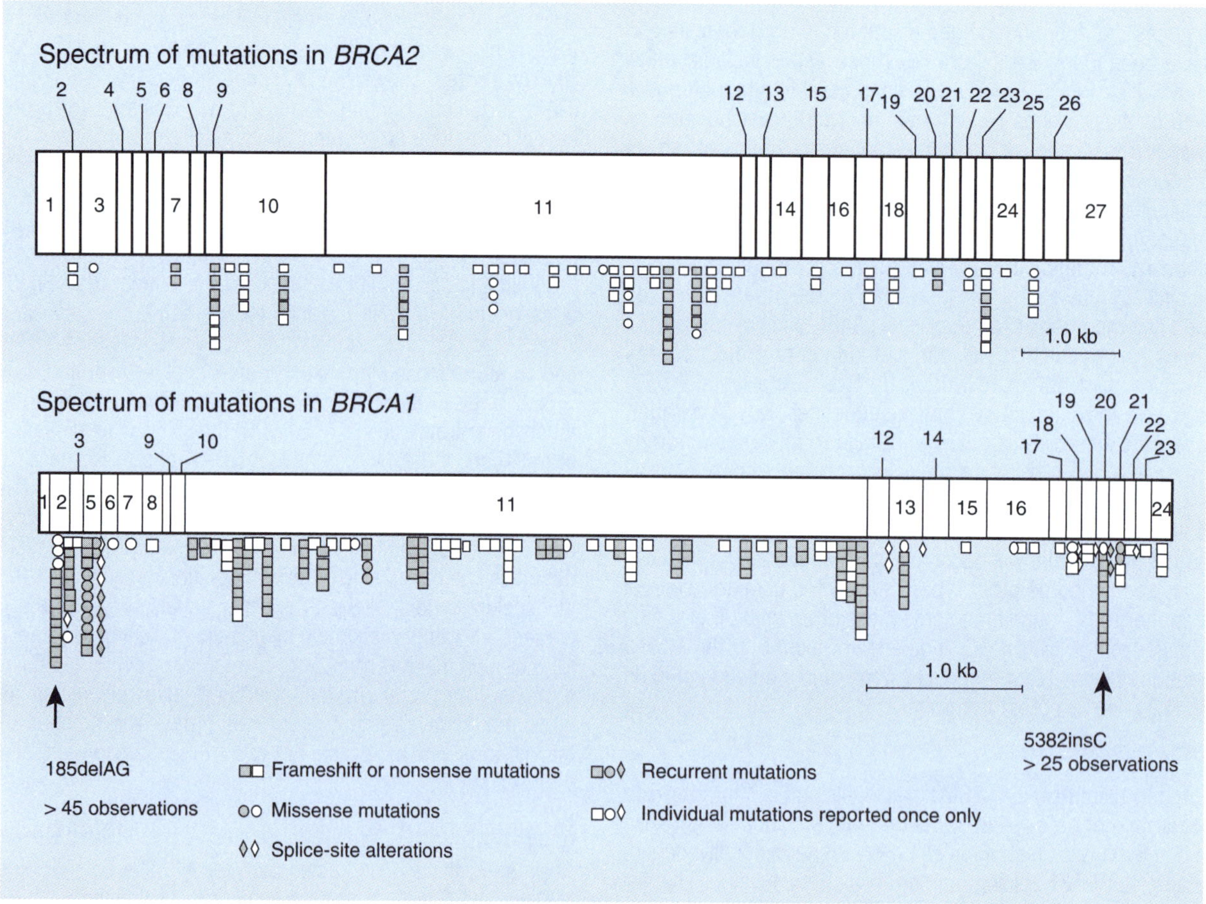

B Common mutations occurring in *BRCA1* and *BRCA2* in Ashkenazi Jewish women

Gene	*Mutation*	*Type of mutation*
BRCA1	185delAG[a]	Frameshift in exon 2
	188del11	Frameshift in exon 2
	5382insC	Frameshift in exon 20
BRCA2	6174delT[b]	Frameshift in exon 11

a Found in 20% of Ashkenazi Jewish women with early-onset breast cancer.
b Found in 8% of Ashkenazi Jewish women with early-onset breast cancer.

OVERVIEW

The autosomal dominantly inherited form of breast and ovarian cancer accounts for approximately 5%–10% of all breast and ovarian cancers found in the United States. Mutations have been identified in two genes, *BRCA1* and *BRCA2*, that appear to account for about two-thirds of the hereditary form of breast cancer. Women who have a *BRCA1* mutation have a lifetime risk of 80% for developing breast cancer and a 50% risk of developing ovarian cancer. A loss of heterozygosity is associated with the disease, which suggests that the *BRCA* genes code for tumor suppressor proteins.

The *BRCA1* and *BRCA2* Genes

The *BRCA1* gene contains 24 exons, spans 100 kb of DNA, and codes for a protein of 1863 amino acids (Figure **A**). The function of the protein is unknown; however, in the C-terminal end of the protein coded for by *BRCA1*, there is a transcriptional transactivation domain that is mutated in families predisposed to breast and ovarian cancer. Consistent with this transcriptional regulatory function is the recent observation that transfecting breast and ovarian cancer cells in vitro with normal *BRCA1* inhibits the growth of the tumor cells. All these characteristics support the role of *BRCA1* as a tumor suppressor gene. The *BRCA2* gene contains 27 exons and codes for a protein of 3418 amino acids, once again with an unknown function. However, recent information has implicated a potential role of the *BRCA2* gene product in DNA repair processes. The *BRCA2* protein interacts with RAD51, a protein involved in repairing double-strand breaks in DNA, which suggests that the *BRCA2* protein may be required for repairing damaged DNA; in its absence, cells accumulate additional mutations that eventually lead to cancer. The *BRCA1* gene product also appears to interact with RAD51, which again suggests a role for this protein in DNA repair processes.

Mutations Associated with *BRCA*

BRCA mutations are rare in the general population, and those that have been identified are spread throughout both genes. A different picture is seen with women in the Ashkenazi Jewish population, where four specific mutations in the *BRCA* genes appear to occur at a high frequency (Figure **B**). Two mutations (185delAG in *BRCA1* and 6174delT in *BRCA2*), one in each of the *BRCA* genes, are found in 1% of the Ashkenazi Jewish population. Female carriers of *BRCA1* mutations have an 80%–90% risk of developing breast cancer and a 40%–50% risk of developing ovarian cancer. The *BRCA1* mutation 185delAG has been found in 20%, and the *BRCA2* mutation 6174delT has been found in 8% of Ashkenazi Jewish women with early-onset breast cancer. The mutation 185delAG occurs in exon 2 and results in a frameshift mutation and the production of a truncated protein. The mutation 6174delT occurs in exon 11 of *BRCA2* and is also a frameshift mutation, leading to premature protein termination. Two other mutations that are common in the Ashkenazi Jewish population are the frameshift mutation 188del11, found in exon 2 of *BRCA1*, and the frameshift mutation 5382insC, found in exon 20 of *BRCA1*. The presence of these common mutations in Jewish women may provide a means of testing and screening this high-risk population.

Issues in Genetic Testing

Clinical testing for human genetic diseases is a highly debated issue. The introduction of commercially available genetic tests for detecting the mutations in the breast cancer genes *BRCA1* and *BRCA2*, which are implicated in 80% of cases of hereditary early-onset breast cancer and ovarian cancer, has raised many questions. In fact, these mutations account for fewer than 10% of the cases of sporadic breast cancer found in the general population. The debate is not whether these prepackaged tests can detect a mutation in these genes, but what the results mean and how they will affect the individual, her family, and her life. Because there is no guarantee that a negative result means a woman will not get breast cancer or that a positive result means she will develop breast cancer, should the test be available for everyone? The individual who plans to take the test should be encouraged to seek genetic counseling. Genetic counselors are trained to explain both the benefits and the disadvantages associated with the test and to counsel the individual on how she should interpret the results and deal with them. As more and more clinical tests for detecting human disease genes become available, decisions about whether to have the test or not will affect family planning, family relationships, and confidentiality of a person's genetic makeup, as well as raise social and legal questions. The federal government has issued some guidelines for dealing with genetic discrimination in the workplace. As of 1995, the Americans with Disabilities Act protects individuals subjected to discrimination on the basis of genetic information relating to illness, disease, or other disorders. In 1996, Congress passed the Health Insurance Portability and Accountability Act, which prohibits a group health insurance plan from using genetic information to establish rules for preventing an individual from being eligible or from continuing to be eligible for insurance. It does not, however, address practices of increasing the cost of insurance or making a blanket exclusion of coverage for a particular condition. At present there are some limited safeguards against discrimination for genetic disease, but there is no firm policy on who has access to genetic information, and additional protection is needed for everyone. The Hereditary Susceptibility Working Group of the National Action Plan on Breast Cancer and the Ethical, Legal, and Social Implications group of the Human Genome Project group have made recommendations to state and federal policymakers that address this issue in more detail in an effort to develop a comprehensive approach to the problem of genetic privacy and genetic discrimination in the workplace.

For more information see Hoffee PA: *IMS: Medical Molecular Genetics*. Fence Creek, pp. 279–280, 356–357.

36 X-Linked Inheritance I

General Concepts—Muscular Dystrophy, an X-Linked Recessive Disorder

Genetics 36

A Typical X-linked recessive pedigree

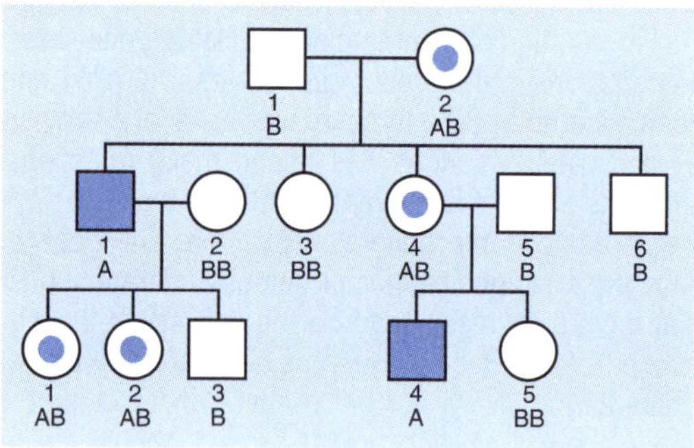

B Typical X-linked dominant pedigree

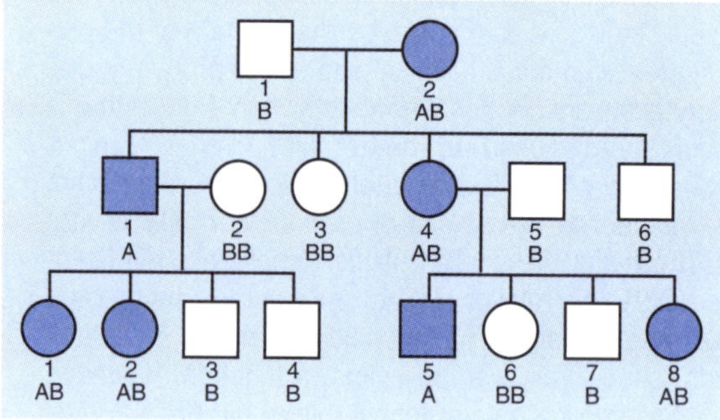

C Duchenne muscular dystrophy-X-linked recessive inheritance

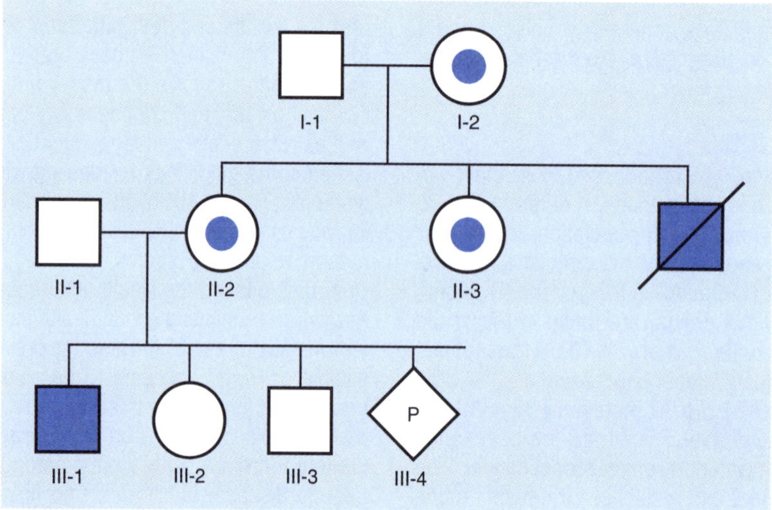

D Examples of X-linked Inherited Disorders

Disorder	*Protein altered*	*Frequency*
Duchenne muscular dystrophy	Dystrophin	1/3500 males
Becker muscular dystrophy	Dystrophin	1/3500
Fragile-X syndrome	FMR1 protein	1/1250 males 1/2000 females
Lesch-Nyhan syndrome	HGPRTase	1/400,000
Spinobulbar muscular atrophy	Androgen receptor	1/100,000
Testicular feminization	Androgen receptor	1/20,000

OVERVIEW

X-linked inheritance has specific characteristics that differ from autosomal inheritance because of the presence of two X chromosomes in the female and only one X chromosome in the male. Both X-linked recessive and X-linked dominant diseases are known; however, most commonly inherited X-linked disorders are recessive.

X-Linked Recessive Pedigrees

A typical X-linked recessive pedigree is shown in Figure **A**, which illustrates the main characteristics associated with X-linked recessive inheritance. (1) There is never father-to-son transmission of X-linked traits, because a father always transmits a copy of his Y chromosome to his son and never the X chromosome. (2) Transmission of X-linked traits from father to daughter is 100%, because the daughter must receive one of her X chromosomes from her father. Therefore, all daughters of an affected male are carriers of the trait. (3) When the mother is a heterozygous carrier, she will transmit the affected allele to 50% of her sons, who will be affected, and to 50% of her daughters, who will be carriers. (4) For most X-linked disorders, the female is generally a heterozygous carrier, and there is a much greater chance that males will be affected with the disorder than females. (5) Affected males in a family are related to each other through carrier females, because males cannot transmit the trait to their sons.

X-Linked Dominant Pedigrees

A typical X-linked dominant pedigree is shown in Figure **B**, which illustrates the characteristics associated with an X-linked dominant disorder. (1) As in X-linked recessive disorders, there is never father-to-son transmission. (2) All daughters of an affected male are affected. (3) Approximately 50% of all offspring of a heterozygous mother will be affected, and 100% of the offspring of a homozygous mother will be affected. (4) Females are twice as likely to be affected as males. X-linked dominant disorders are rare in the population and are often difficult to distinguish from an autosomal dominant disorder simply on the basis of pedigree analysis. Once the candidate gene for the disease has been mapped, however, a clear distinction can be made between autosomal dominant and X-linked dominant patterns of inheritance.

Duchenne Muscular Dystrophy (DMD) and Becker Muscular Dystrophy (BMD)

DMD is the most severe and most common form of muscular dystrophy and affects one in 3500 males. The disease is generally fatal and currently has no treatment. Because of the excessively large size of the DMD gene, nearly 2.5 million bp, many cases of the disease are due to new mutations without a previous family history of the disease. The high mutation rate (10^{-4} in the sperm and the egg), which is at least 10–100 times that found for most genetic diseases, implies that every male with a sperm production of 10^7–10^8 a day will produce at least one sperm every day with a mutation in the DMD gene. Because the disease is inherited as an X-linked recessive trait, as a rule only males are affected with the disease, and females are carriers (Figure **C**). There is no specificity to the disease; all populations as well as many animal species can be affected. Boys who inherit the disease gene are asymptomatic at birth and appear to show few if any symptoms until 4–5 years of age. By the age of 2 years, they may show some calf hypertrophy with noticeably large muscles, but this abnormality often goes undetected by the parents. By the time the children are in first grade, they begin to show some muscle weakness. A blood test at this age shows that creatine kinase levels are extremely high (50–100 times the normal level), which is an indication of muscle destruction. By age 11 years, the boys are confined to wheelchairs, and by age 15 years, they have little muscle left. The patients generally die of respiratory failure by age 20 years if not ventilated. Another form of muscular dystrophy, BMD, has a similar phenotype but is much milder. Patients with BMD are still walking at 16 years and can live for many years. Both DMD and BMD are a result of mutations in the same X-linked gene. Before 1987, there was little information about the molecular basis for the muscle wasting in DMD. The muscle from DMD patients showed muscle fibers that appeared to die and then grow back, only to die again and finally be replaced by scar tissue. Knowing that DMD was an X-linked trait simplified the isolation and mapping of the DMD gene by positional cloning (Chapter 51) and resulted in the identification of the protein product of the DMD gene as dystrophin. Dystrophin is present in muscle biopsies of normal patients but completely absent in all patients with DMD. The level of dystrophin is also reduced in patients with BMD. The difference in the severity of DMD compared with BMD is apparently related to the amount of dystrophin present in the muscles of the patient. Analysis of muscle tissue using antibodies to dystrophin indicate that dystrophin is present in BMD patients, but its molecular weight is abnormal. Patients with DMD, on the other hand, have no dystrophin protein in their muscle at all. Although both forms of the disease result from deletions occurring in the same gene, the type of deletion, rather than the extent of the deletion, determines the severity of the disease. Following a deletion, correct splicing of the remaining exons in the RNA occurs in BMD patients and allows them to produce a modified, partially functioning dystrophin molecule. In DMD patients, however, the deletion occurs in a manner that results in the remaining exons being spliced out of frame. This aberrant splicing gives a mRNA that is translated out of frame, resulting in the encounter of a termination codon in the middle of the mRNA and premature termination of protein synthesis. When this occurs, the result is a complete absence of the dystrophin protein and a severe clinical phenotype.

Examples of X-Linked Inherited Disorders

Additional examples of X-linked diseases are listed in Figure **D** and discussed in Chapters 37 and 58.

For more information see Hoffee PA: *IMS: Medical Molecular Genetics.* Fence Creek, pp. 150–156, 242–243, 282–284.

37 X-Linked Inheritance II

The Fragile X Syndrome

A 5′ end of the *FMR1* gene

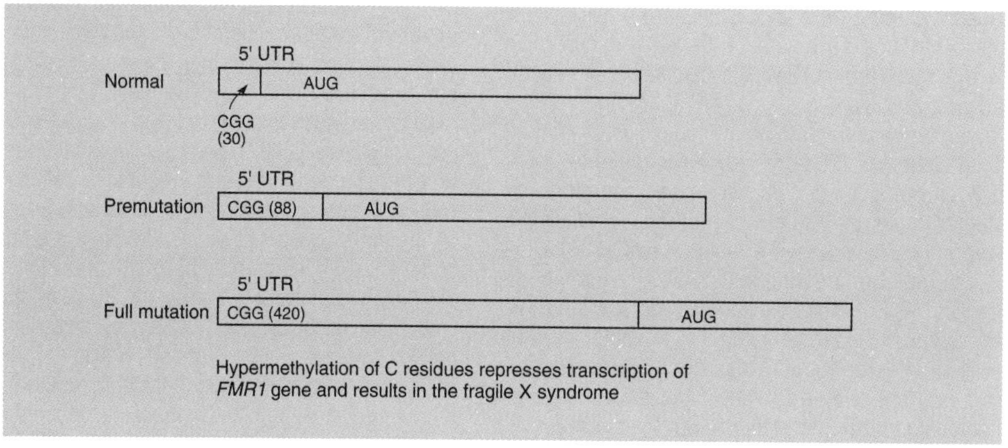

B PCR analysis for fragile X mutation

C Pedigree for a family with fragile X syndrome

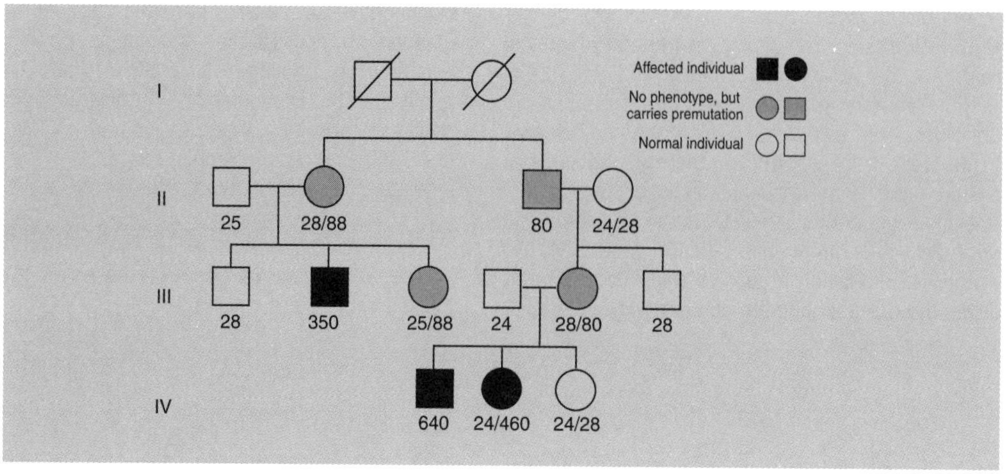

OVERVIEW

The most common inherited form of mental retardation currently known is the fragile X syndrome. This disorder, which affects 1 in 1250 males and 1 in 2000 females, is characterized cytogenetically by the presence at the end of the long arm of the X chromosome of a non-staining region. This cytogenetic abnormality is seen when cells taken from individuals with the fragile X syndrome are cultured in a medium deficient in folic acid and thymidine. Metaphase chromosomes prepared from cells grown under these culture conditions will show a lack of staining and an apparent breakage on the X chromosome at Xq27.3 (i.e., a fragile X). Individuals with fragile X syndrome present with both physical and nonphysical symptoms. Physical symptoms include large ears; a long, narrow face with a prominent forehead and a square chin; and increased testicular volume (macroorchidism) in males. Nonphysical symptoms include mental retardation (usually the most common symptom), developmental delay, learning disabilities, and autistic-like features. Females with the fragile X syndrome are less severely affected than males, having less mental retardation, fewer behavior problems, and less obvious physical symptoms.

Molecular Basis for Fragile X Syndrome

With the cloning of the gene (*FMR1*) responsible for the fragile X syndrome, the molecular basis for the disease and its unusual inheritance pattern have become much clearer. The *FMR1* gene spans 38 kb and consists of 17 exons. In the noncoding 5′ end of the first exon of the gene is a trinucleotide repeat sequence, CGG, which is present in normal individuals in a range of 6–50 copies, with 30 copies of the repeat being the most common (Figure **A**). Expansion of this trinucleotide repeat sequence is associated with the disease. When the repeat lengths are expanded to 60–200 copies, the premutation, or carrier state, is seen in both males and females. The carrier state is not usually associated with a clinical phenotype. However, the premutation expands with a high probability in carrier females to give a full mutation of 230–1000 copies of the CGG repeat and the fragile X syndrome. Most males with the full mutation and approximately 50% of females heterozygous for the full mutation show a clinical phenotype. Within the *FMR1* gene of affected individuals, the C residues present in the expanded CGG repeats and nearby CG repeats are hypermethylated. Hypermethylation interferes with normal transcription of the *FMR1* gene and results in the absence of *FMR1* RNA. In a normal individual, the gene is unmethylated and produces normal amounts of RNA, with the highest expression of the gene occurring in the brain and testes. The precursor RNA undergoes alternative splicing, and multiple FMR1 protein isoforms found in the cytoplasm contain RNA-binding domains, suggesting a role for the proteins in RNA metabolism. Normal *FMR1* proteins appear to be associated with the 60S ribosomal subunit and may be involved in translation. The absence of the proteins in an individual with the fragile X syndrome potentially interferes with normal translation, especially in the brain and testes, to give the observed phenotype. Additionally, heavy methylation of the large expansion region may alter DNA replication timing and/or chromatin structure in such a way as to affect the condensing of the chromatin at mitosis, which is reflected in the cytogenetic fragile X structure.

Molecular Diagnosis for the Fragile X Syndrome

Direct DNA analysis can be used to diagnose the syndrome and is more reliable than the cytogenetic analysis for the fragile X site. DNA from a blood sample is used either in a Southern blot analysis or in PCR analysis to determine the length of the CGG repeat at the 5′ end of the gene. Hybridization of the PCR products to a CGG oligonucleotide probe detects different-size repeat fragments that fall into the normal, the premutation, or the full mutation range, as diagramed in Figure **B**. Males with only one X chromosome have only one PCR product, whereas females with two X chromosomes have two PCR products. Prenatal diagnosis, particularly for a carrier of the fragile X mutation, can be performed by either chorionic villus sampling or amniocentesis.

Inheritance Pattern for Fragile X Syndrome

The fragile X syndrome is inherited as an X-linked disorder but does not follow a simple X-linked recessive or X-linked dominant pattern of inheritance, as indicated by the pedigree shown in Figure **C**. All males with the full mutation show clinical symptoms; however, carrier males, known as transmitting males, exist who carry the premutation and are usually phenotypically normal. A carrier male (II-3) transmits the premutation to all of his daughters, and they all become carriers. However, transmission of the premutation from a father to his daughter does not usually result in expansion to the full mutation or a clinical phenotype in the daughter. Expansion of the premutation to the full mutation occurs predominantly in maternal transmission, in which case the clinical syndrome is seen in both sons and daughters of carrier females (III-5). For reasons still unexplained, the expansion of the CGG repeat from the premutation to the full mutation state occurs in maternal meiosis and not in paternal meiosis. Thus, females with a full mutation and a clinical phenotype occur only when the expansion is transmitted from the mother, not when it is transmitted from the father. Carrier females can be of two types, as indicated in Figure **B**. They can carry a normal and a premutation allele or a normal and a full mutation allele. Unlike most X-linked recessive disorders 50% of females heterozygous for the full mutation allele show clinical symptoms, although they are generally less severe than in males. In families with the fragile X syndrome, the disease gets worse in successive maternally derived generations, a phenomenon known as genetic anticipation. This characteristic is related to the expansion of the repeats from a premutation to a full mutation in carrier females and transmission of the full mutation to both sons and daughters.

For more information see Hoffee PA: *IMS: Medical Molecular Genetics*. Fence Creek, pp. 282–284.

38 Mitochondrial Inheritance

Diseases Associated with Mutations in Mitochondrial DNA

A Protein complexes involved in oxidative phosphorylation

Complex	*No. mitochondrial subunits*	*Mitochondrial genes*
I. NADH → coenzyme Q	7	*ND1—6*
II. Succinate coenzyme A reductase	0	—
III. Coenzyme Q → cytochrome C	1	*Cytochrome b*
IV. Cytochrome C → oxygen	3	*CO I—III*
V. ATP synthetase (ADP + P_i →ATP)	2	*ATPase 6; ATPase 8*

Note: ADP = adenosine diphosphate; ATP = adenosine triphosphate; NADH = reduced nicotinamide dehydrogenase; P_i = inorganic phosphate

B Mitochondrial genome

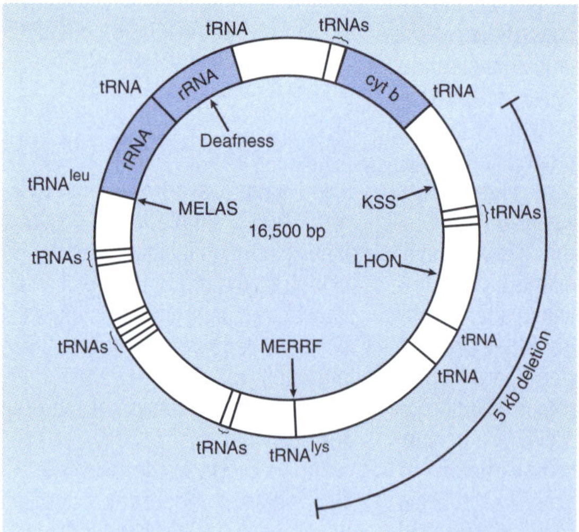

C Diseases associated with mutations in mitochondrial DNA

Disease	*Common mutation associated with the disease*
LHON	Missense mutation in ND4 (LHON 11778); defects in complex I
MELAS	Base substitution mutation in tRNA-leucine (MELAS 3243)
MERRF	Base substitution mutation in tRNA-lysine (MERRF 8344)
Deafness	Base substitution mutation in 12s rRNA (12srRNA 1555)

Note: LHON = Leber's hereditary optic neuropathy; MELAS = mitochondrial myopathy, encephalomyopathy, lactic acidosis and stroke-like episodes; MERRF = myoclonic epilepsy with ragged red fibers

D Inheritance pattern of mitochondrial diseases

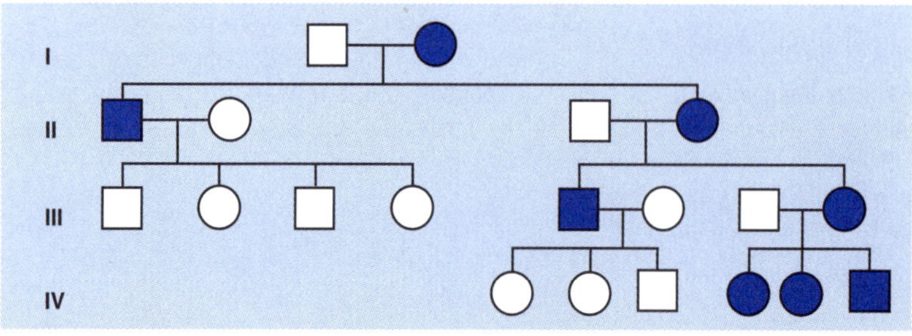

OVERVIEW

Crucial to the normal life of a cell is the production of energy in the form of ATP, a process carried out in the mitochondria. These organelles, present in one hundred copies per cell, carry out oxidative phosphorylation, using enzyme complexes located in their inner membranes to provide cells with their major source of ATP.

Protein Complexes Involved in Oxidative Phosphorylation

The five protein complexes and the mitochondrial genes involved in oxidative phosphorylation are listed in Figure **A**. Complexes I–V are made up of many subunits, only a few of which are coded for by mitochondrial genes, the majority being coded for by nuclear genes. Mutations in mitochondrial genes that alter the structure of a subunit or the structure of the tRNAs or rRNAs necessary to synthesize a subunit can lead to mitochondrial disorders.

Mitochondrial Genome

The mitochondrial DNA, a closed circular double-stranded DNA molecule 16.5 kb in size, codes for 22 tRNA molecules, 2 rRNA molecules, and 13 protein subunits that are part of the complexes involved in the respiratory chain (Figure **B**). Unlike nuclear DNA, mitochondrial DNA has no introns, is not complexed to histones, and appears not to be subject to the various repair processes that take place in the nucleus. These unique characteristics of the mitochondrial DNA contribute to the nearly 10-fold increase in its mutation rate compared with nuclear DNA.

Diseases Associated with Mutations in Mitochondrial DNA

Every cell in the body contains hundreds of mitochondria, each mitochondrion containing 2–10 molecules of DNA. Within a single cell, and even within a single mitochondrion, there can exist DNA molecules that differ from one another in base-pair sequence. The term *homoplasmy* refers to a cell that contains all normal or all mutant mitochondrial DNA. The term *heteroplasmy* refers to the presence of both mutant and normal mitochondrial DNA in the same cell. A 5-kb region of the mitochondria DNA (noted in Figure **B**) is referred to as the common deletion area. Deletions that result in the syndrome known as Kearns-Sayre syndrome (KSS) occur in this region and give rise to defects in complexes I, III, and IV. The onset of symptoms occurs before 20 years of age and the disease is characterized by progressive external ophthalmoplegia, pigmentary degeneration of the retina, and heart block, or cerebellar syndrome. This disease is heteroplasmic, and most cases appear to result from spontaneous deletions occurring in the mitochondrial DNA during embryogenesis rather than from maternal transmission. The entire base sequence of the mitochondrial DNA is known, allowing the identification of single-base pair substitutions that are associated with specific disease phenotypes. Some of the most common are listed in Figure **C** and shown in Figure **B**. Leber's hereditary optic neuropathy (LHON) is one of the most commonly inherited causes of blindness in young males under the age of 25. The disease is maternally inherited and shows increased penetrance in males (3–4 times as in females with the same mutation). The blindness, which is of sudden onset and progresses rapidly, is due to degeneration of the optic nerve. A mutation (LHON 11778) in the gene coding for ND4, a subunit found in complex I, leads to LHON. Many individuals with this LHON mutation are homoplasmic; however, heteroplasmic individuals also exist. Recovery of vision occurs in fewer than 5% of individuals whose blindness results from this mutation. Mitochondrial myopathy, encephalopathy, lactic acidosis, and stroke-like episodes (MELAS) is one of the mitochondrial encephalomyopathies that affects the nervous system, the eye, and some somatic organs. This is a progressive disease; the first symptoms occur between 5 and 15 years of age. Symptoms include ataxia tremors and myoclonic jerks. MELAS is distinguished from other similar diseases by the occurrence of stroke-like episodes and may lead to cortical blindness and often dementia. Nearly 80% of individuals with MELAS are heteroplasmic, with the mutant mitochondrial DNA having a specific A-to-G substitution at bp 3243 in the gene that codes for mitochondrial tRNA-leu. Individuals with myoclonic epilepsy with ragged red fibers (MERRF) have muscle fibers with too many mitochondria. Symptoms of MERRF appear between 5 and 12 years of age and are similar to those of MELAS. However, no stroke-like symptoms are found in MERRF. The disease, which is the result of a mutation in the gene coding for mitochondrial tRNA-lys, has variable severity, is usually heteroplasmic, and is characterized by progressive myoclonic epilepsy. Patients commonly have seizures and develop dementia, hearing loss, and optic atrophy. Nonsyndromic deafness shows a maternal inheritance pattern, which suggests that this disease may result from mitochondrial mutations. This disease is characterized by irreversible hearing loss that is associated with the use of aminoglycoside antibiotics and is a result of a mutation in the mitochondrial gene coding for the 12s rRNA.

Inheritance Pattern

Mitochondrial disorders are maternally inherited because the egg cell and not the sperm is responsible for transmitting mitochondrial DNA molecules to the offspring. The pedigree in Figure **D** illustrates the main characteristic of mitochondrial inheritance: an affected father does not transmit a mitochondrial disease to his children, whereas an affected mother transmits the disease to all of her children. The severity and variability of the clinical phenotype in mitochondrial diseases is often related to the ratio of mutant to normal mitochondrial DNA in a cell. Furthermore, some cells in the body require more energy than others (e.g., muscle, eye), resulting in thresholds at which a certain level of mutant mitochondrial DNA must be present before a cellular dysfunction occurs. This may be the reason why many of the mitochondrial disorders are chronic, late-onset diseases that involve environmental factors.

For more information see Hoffee PA: *IMS: Medical Molecular Genetics*. Fence Creek, pp. 284–287.

39 Population Genetics

Hardy-Weinberg Equilibrium

A Allele frequency

Parents	*A*	*B*
A	*AA*	*AB*
B	*BA*	*BB*

P(*A*) = 0.6 P(*B*) = 0.4

p + q = 1.0

B Genotype frequency

$P(AA) = p^2$ $\qquad P(AA) = p^2 = 0.36$

$P(AB) = 2pq$ $\qquad P(AB) = 2pq = 0.48$

$P(BB) = q^2$ $\qquad P(BB) = q^2 = 0.16$

$p^2 + 2pq + q^2 = 1.0$ $\qquad 0.36 + 0.48 + 0.16 = 1.0$

C Pedigree illustrating identity by descent

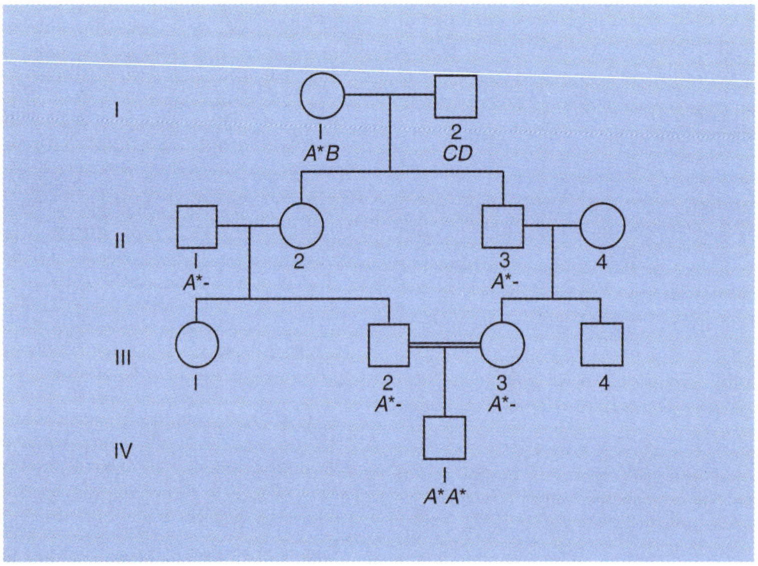

D Coefficients of relatedness (R) and inbreeding (F)

Relatives	R	F
Siblings	1/2	1/4
Uncle or aunt; niece or nephew	1/4	1/8
First cousins	1/8	1/16
Second cousins	1/32	1/64

OVERVIEW

Studies of population genetics use statistical analysis to estimate the allelic and genotypic frequencies within a population and to evaluate factors that alter genetic frequencies with time. These investigations are important for estimating recurrence risks for inherited human diseases as well as for studying genetic patterns in different populations.

Allelic Frequencies

For each human gene located on an autosomal chromosome, there are two alleles, one on each of the homologous chromosomes. If we assume that the two alleles are *A* and *B*, then in a given population, the probability (P) of the *A* allele occurring in a population is p, and the P of the *B* allele occurring in a population is q. The sum of the two probabilities, P(*A*) and P(*B*), must add up to 1.0. Therefore, the allelic frequencies are $p + q = 1$ or $q = 1 - p$ (Figure **A**).

Genotypic and Phenotypic Frequencies

If there are two alleles, *A* and *B*, then according to mendelian predictions there will be three genotypes; *AA*, *AB*, and *BB*. If a given population has random mating, the Hardy-Weinberg equilibrium can be used to calculate the genotypic frequencies from the allelic frequencies. When a population is in Hardy-Weinberg equilibrium, in which $P(AA) = p^2$, $P(AB) = 2pq$, and $P(BB) = q^2$, then $p^2 + 2pq + q^2 = 1.0$ (Figure **B**). If the frequency of a particular allele is known, then the frequency of the genotype present in the population can be determined. For example, if the frequency of the *A* allele (p) is 0.6, then the q, or *B*, allele must equal 0.4. If the population is in Hardy-Weinberg equilibrium, then p^2 [P(*AA*)] is 0.36, 2pq [P(*AB*)] is 0.48, and q^2 [P(*BB*)] is 0.16, or $0.36 + 0.48 + 0.16 = 1.0$ (see Figure **B**). Similarly, the genotypic frequencies, which one observes, can be used to calculate the allelic frequencies. For example, if the proportion of the *AA* genotype is 0.64, the *AB* genotype is 0.32, and the *BB* genotype is 0.04, then the sum of these three is 1.0. If they are consistent with the Hardy-Weinberg equilibrium, $P(AA) = p^2$ and $P(BB) = q^2$; thus, $p = 0.8$ and $q = 0.2$. Once again, $p + q$ must equal 1. If the genotypes at a particular locus can be distinguished, then the phenotypic and genotypic frequencies are the same. However, in inherited human diseases, dominant-recessive relationships between two alleles preclude distinguishing two genotypes that result in the same phenotype. For example, if *A* is a dominant allele and *B* is a recessive allele, then both genotypes, *AA* and *AB*, will have the same phenotype. For calculation of the phenotypic frequencies from the genotypic frequencies, information is needed about the mode of inheritance of a locus and the penetrance of each genotype. Penetrance is defined as the probability that an individual with a particular genotype will have an affected phenotype. Some inherited human diseases show incomplete or reduced penetrance, in which case an individual may carry the disease allele and not express it, or it may be expressed to a lesser degree. Diseases that have a late onset or that are subject to environmental influences can show reduced or incomplete penetrance. Other diseases are fully penetrant, and every individual carrying the disease allele expresses the affected phenotype. The following is an example of using phenotypic frequencies to infer genotypic or allelic frequencies. In a population in Hardy-Weinberg equilibrium, 0.01% of individuals are affected with an autosomal recessively inherited disorder that is fully penetrant. Therefore, P (the affected recessive allele) = 0.0001 = q^2 (*BB*), or q = .01. Since p + q must equal 1.0, then p, or the dominant allele, is equal to 0.99.

Factors That Modify Genotypic Frequencies

Among different populations, the genotypic frequencies at most genetic loci differ. Several factors are known to influence the genotypic frequencies within a population. Nonrandom mating, in particular inbreeding (i.e., mating with the descendant of a common ancestor) results in an increase in homozygosity within a population. Homozygosity caused by inbreeding shows identity by descent, which occurs when a single DNA sequence is inherited from the same ancestor, as indicated in Figure **C**. In this example, individual IV-1 has received an *A** from his mother and an *A** from his father. The *A** alleles are identity by descent because both parents received the *A** that originated in the grandmother (I-1). The parents are first cousins; therefore, individual IV-1 is the offspring of a consanguineous marriage (i.e., a marriage of two individuals with at least one ancestor in common, usually no further removed than a great-grandparent) and is said to be inbred. Inbreeding is defined as the mating between two consanguineous individuals.

Coefficient of Relatedness and Inbreeding

The coefficient of relatedness (R) gives an estimate of how closely related two people are. R is equal to the probability that two individuals share an allele by identity by descent. The coefficient of inbreeding (F) estimates how closely related an individual's parents are. F is equal to the probability that an individual is homozygous and has the alleles by identity by descent. R equals 2F; therefore, one can calculate the values of F from R for matings between individuals of various degrees of relationships, as shown in Figure **D**. For example, a marriage between siblings has an R value of ½; a marriage between first cousins has a value of $^1/_8$.The coefficients of inbreeding are then ¼ and $^1/_{16}$ respectively.

Clinical Significance

Homozygosity is increased with inbreeding within a population and can lead to an increase in the number of individuals affected with an autosomal recessive disorder that requires a defective allele to be inherited from each parent. Some populations are at higher risk for a particular disease because of the increased number of individuals who carry the defective allele and the probability that mating within a particular population is increased.

For more information see Hoffee PA: *IMS: Medical Molecular Genetics*. Fence Creek, pp. 230–237.

40 Recurrence Risk Estimations

A Bayesian probability calculations

These calculations are used to determine the relative probabilities of two mutually exclusive alternatives. For each alternative, four components are calculated:

1. **Prior probability,** the initial probability without regard for complications or for additional information
2. **Conditional probability,** the probability of the additional information under each alternative
3. **Joint probability,** the product of the prior and conditional probabilities
4. **Posterior probability,** the joint probability of each alternative divided by the sum of the two joint probabilities

B Pedigree of a family with Huntington disease

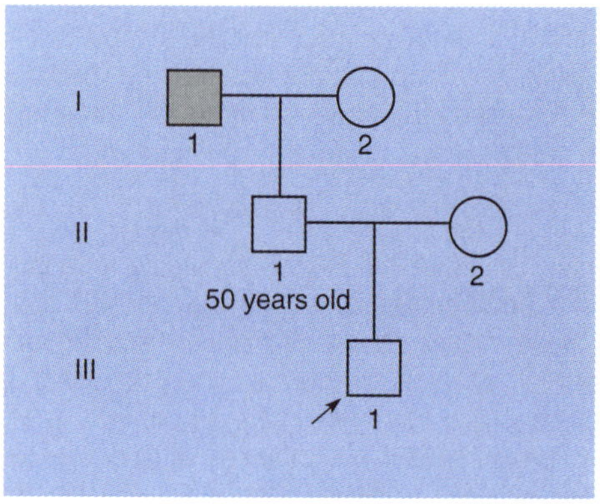

C Example of a Bayesian calculation

Probability	Carrier	Not a Carrier
Prior	$1/2$	$1/2$
Conditional	P (not affected by 50 years) = $1/5$	1
Joint	$1/2 \times 1/5 = 1/10$	$1/2 \times 1 = 1/2$
Posterior	$(1/10) \div (1/10 + 1/2) = 1/6$	$(1/2) \div (1/10 + 1/2) = 5/6$

Note. Calculated are the relative probabilities that individual II-1 in Figure **B** is either a carrier for Huntington disease or not a carrier.

D Pedigree of a family with cystic fibrosis

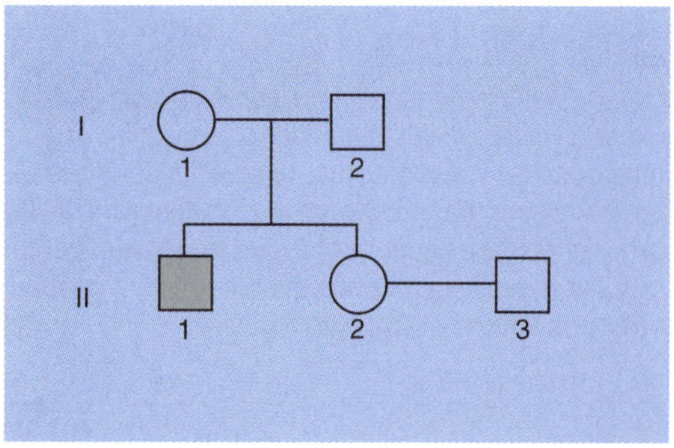

OVERVIEW

Recurrence risk estimations are statistical calculations that allow a medical geneticist to estimate the risk that a particular disorder will recur in a relative of an affected person. For example, a couple with one child who has a genetic disorder may want to know what the chances are that their second child will be born with the same disorder. An individual who has a relative with a particular disorder may want to know if he or she is at risk for having the same disorder. In this analysis, arriving at an accurate diagnosis for the disorder in question is important. This may involve taking detailed family histories, examining affected individuals within a family, and performing cytogenetic analysis and DNA testing.

Conditions of Known Etiology

When the basis of a disorder is known, the analysis is carried out by using probabilities that are based on mendelian segregation patterns (discussed in Chapter 25). If the inheritance pattern is clear (i.e., an autosomal dominant disorder that shows complete penetrance), then the recurrence risk is based on straight mendelian predictions. In this case, the risk of an affected parent having an affected child is 50%, and for an unaffected parent, the risk is essentially zero (or equal to that of the general population). Unfortunately, for many inherited human disorders, the inheritance pattern is not straightforward. Factors such as reduced penetrance, variable expressivity, and late onset hinder the estimations for dominant disorders. Recessive disorders offer a further complication in that the genotype of unaffected individuals must be determined, because an unaffected individual for an autosomal recessive disorder can be heterozygous for the defective allele. Additionally, current methodologies allow the results from molecular studies to be incorporated into the risk calculations. Bayesian probability calculations can be used to incorporate all these factors into the risk determination. Bayesian probability calculations are used to determine the relative probabilities of two mutually exclusive alternatives, A and B. For each of the alternatives, four values are determined: the prior probability, the conditional probability, the joint probability, and the posterior probability (Figure **A**).

Using Bayesian Calculations to Determine Risk Recurrence

Huntington disease, a late-onset, progressive, fatal disease is one of the trinucleotide expansion disorders (Chapter 30). Because of the late onset of the disease (beginning at approximately 36–40 years of age) transmission of the disease to offspring occurs before the disease has been diagnosed. A pedigree (Figure **B**) is for a 20-year-old male, III-1, whose grandfather died with Huntington disease. The young man's father is 50 years old and remains unaffected by the disease. What is the risk that the young man has inherited Huntington disease? Initially, his risk would be $^1/_4$, or equal to $^1/_2$ that II-1 is a carrier times $^1/_2$ that the defective allele was transmitted to III-1. Because 80% of individuals carrying the defective allele are affected by age 50, and the father has reached 50 years of age without being affected, this factor can be taken into account when the recurrence risk for III-1 is determined. The first assumption is that the father (II-1) is either A, a carrier, or B, not a carrier, since III-1 will be affected only if his father is a carrier. The calculation is summarized in Figure **C**. The prior probability is $^1/_2$ that II-1 received the defective allele from the affected I-1. The conditional probability is the probability that a carrier will not be affected at age 50, or 20%. If he is not a carrier, the conditional probability that II-1 will be unaffected is 100%. The joint probabilities are the product of the prior and conditional probabilities for each alternative. The posterior probabilities are each joint probability divided by the sum of the two joint probabilities, or $^1/_6$ that the father (II-1) is a carrier and $^5/_6$ that he is not a carrier. We now estimate the risk that the son, III-1, has the defective allele as follows: P(II-1 is a carrier) x P (the allele is transmitted to III-1) = $^1/_6 \times {}^1/_2$, or $^1/_{12}$. Using Bayesian calculations, the recurrence risk for III-1 is reduced from 25% to 8.3%.

Recurrence Risk for Sibling Carriers

Recurrence risk estimations for autosomal recessive disorders have the further complication that the genotype is unknown for unaffected individuals. Figure **D** shows a pedigree for a family with CF. Individuals II-2 and II-3 want to know what the risk is that their children will be affected with CF, which affected II-1, the brother of II-2. In order for II-2 to be affected, both I-1 and I-2 must be heterozygous for the CF allele. For an autosomal recessive disorder, mendelian segregation patterns predict that the expected offspring of two *AB* unaffected heterozygous individuals will be $^1/_4$ *AA*, $^1/_2$ *AB*, and $^1/_4$ *BB*, or $^1/_4$ affected (*BB*) and $^3/_4$ unaffected (*AA*, *AB*). Therefore, the unaffected offspring are $^1/_3$ *AA* and $^2/_3$ *AB*. Because II-2 is unaffected, the probability that she is heterozygous (*AB*) is $^2/_3$. For the couple (II-2 and II-3) to have an affected child, II-3 must also be a carrier. Because he has no family history of CF, the carrier state of II-3 is the same as that of the general population. Thus, the probability that both II-2 and II-3 are carriers equals $^2/_3$ times the population heterozygote frequency ($^1/_{22}$ for CF), or ($^2/_3 \times {}^1/_{22}$), or $^1/_{33}$. The probability that II-2 and II-3 will have a child with CF is the probability that both are carriers ($^1/_{33}$) times the probability that they will have an affected offspring ($^1/_4$), or $^1/_{132}$.

Clinical Significance

With a substantial increase in the number of human genetic diseases that can be diagnosed at the molecular level, new ways are becoming available to determine whether an unaffected individual is heterozygous for a particular autosomal recessive disorder. Additionally, more diseases can now be diagnosed prenatally, giving potential parents more information to allow them to make informed decisions about family planning and to know the risk that their children will be born with a genetic disorder. In the absence of such information, recurrence risk estimations can give the individual a more accurate determination of the probability that they are carriers or are at risk for a particular human disease.

For more information see Hoffee PA: *IMS: Medical Molecular Genetics*. Fence Creek, pp. 244–250.

PART II: QUESTIONS

Directions: For each of the following questions, choose the **one best** answer.

Questions 1–4

1. Point mutations that result in change-of-function (dominant disorders) of the protein are most often of which of the following type?

(A) Nonsense codon in the middle of the coding region of the gene

(B) Missense mutation that leads to an amino acid substitution in the protein

(C) Mutation that leads to errors in RNA splicing

(D) Mutation that reduces the transcription rate of the protein

(E) Mutation that leads to an error in the translation of the protein

2. Which one of the following disorders is caused by a point mutation within a structural gene (i.e., coding for a protein subunit) in the mitochondrial DNA?

(A) Kearns-Sayre syndrome (KSS)

(B) Myclonic epilepsy and ragged red fibers (MERRF)

(C) Leber's hereditary optic neuropathy (LHON)

(D) Familial hypercholesterolemia

(E) Myopathy, encephalopathy, lactic acidosis, and stroke-like episodes (MELAS)

3. Consider a locus with 2 alleles (*C* and *D*) in a population that is in Hardy-Weinberg equilibrium. If the frequency of the *CC* genotype is 0.36, what is the allele frequency of the *D* allele?

(A) 0.20

(B) 0.24

(C) 0.36

(D) 0.40

(E) 0.48

4. Assume a locus has two alleles, *1* and *2*. The following table lists the genotypic proportions at this locus for five populations (A–E). Which of the populations is in Hardy-Weinberg equilibrium?

	1,1	*1,2*	*2,2*
(A)	0.16	0.59	0.25
(B)	0.25	0.65	0.10
(C)	0.00	1.00	0.00
(D)	0.49	0.42	0.09
(E)	0.32	0.64	0.04

Directions: The group of questions below consist of lettered choices followed by several numbered items. For each numbered item, select the appropriate lettered option with which it is most closely associated. Each lettered option may be used once, more than once, or not at all.

Questions 5–12

The following questions refer to the four pedigree diagrams below, each of which depicts the family pattern of a fully penetrant genetic disorder. Match each mode of inheritance listed below with the correct pedigree.

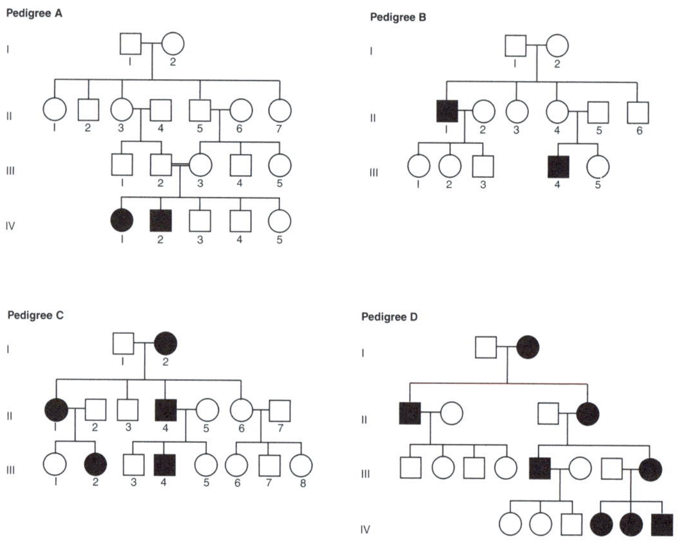

5. Autosomal recessive

6. Autosomal dominant

7. X-linked recessive

8. Mitochondrial inheritance

9. The inheritance pattern for phenylketonuria (PKU)

10. The inheritance pattern for Duchenne muscular dystrophy (DMD)

11. The inheritance pattern for Li-Fraumeni syndrome

12. The inheritance pattern for myoclonic epilepsy and ragged red fibers (MERRF)

Questions 13–17

Match each of the following descriptions with the human genetic disorder it best describes.

(A) Gaucher disease

(B) Tay-Sachs disease

(C) Type I osteogenesis imperfecta

(D) Type II osteogenesis imperfecta

(E) Neurofibromatosis 1 (NF-1)

(F) Neurofibromatosis 2 (NF-2)

(G) Fragile X syndrome

13. This autosomal dominant disorder is a result of a missense mutation in the type I collagen gene and leads to a lethal form of the disease in infants.

14. This autosomal recessive disorder is a result of a defect in glycolipid metabolism and is common among Ashkenazi Jewish populations. Four common mutations are associated with the disease: L44P, N370S, IVS2+1, and 84GG.

15. This disorder is the most common inherited form of mental retardation known and affects 1 in 1250 males and 1 in 2000 females.

16. This disorder is diagnosed by the appearance of café-au-lait spots (at least six spots larger than 15 mm in size) and the presence of at least two Lisch nodules on the iris.

17. This disorder is a result of a mutation in a protein that is involved in the inactivation of a GTP binding protein known as Ras.

Questions 18–22
Match each of the following descriptions with the human genetic disorder it best describes.

(A) Cystic fibrosis (CF)

(B) Familial hypercholesterolemia

(C) Familial adenomatous polyposis (FAP)

(D) Hereditary nonpolyposis colorectal cancer (HNPCC)

(E) Myotonic dystrophy

(F) Duchenne muscular dystrophy (DMD)

(G) Becker muscular dystrophy (BMD)

18. This disorder has been shown to be a result of mutations occurring in the genes involved in the MMR.

19. This autosomal dominant disorder in the heterozygous state has an incidence of 1 in 500 individuals and accounts for 5% of the coronary artery diseases in the Western world.

20. A specific mutation, known as Δ F508, is associated with 70% of the mutant alleles that lead to this autosomal recessive disorder.

21. The protein defective in this disorder is normally involved in the down-regulation of the β-catenin transcription function.

22. This trinucleotide expansion disorder is associated with CTG repeats in the noncoding 3′ end of the last exon of a gene coding for a cAMP-dependent kinase.

PART II: ANSWERS AND EXPLANATIONS

1. **The answer is B.**

In dominantly inherited disorders, a mutant protein is made that has a dominant effect over the normal protein. A missense mutation that leads to an amino acid change will produce a mutant protein with a single amino acid change from the normal protein. To produce a dominant phenotype, that altered protein must be stable, must be synthesized at a normal rate, and must interfere with the normal gene product. The other mutations generally result in the absence of a mutant protein or a decrease in the level of the normal protein. In the absence of a dominant mutant protein, the normal protein produced from the normal allele is generally sufficient to preclude a clinical phenotype.

2. **The answer is C.**

LHON is caused by a mutation in the ND4 structural gene that leads to a defect in oxidative phosphorylation complex I. Both MERRF and MELAS are caused by mutations in mitochondrial transfer RNA genes. KSS is caused by a deletion in the mitochondrial DNA. Familial hypercholesterolemia is not associated with a defect in mitochondrial DNA and is caused by a mutation in the LDL receptor.

3. **The answer is D.**

If the frequency of the *CC* genotype is 0.36, then the frequency of the *C* allele is 0.6. When a population is in Hardy-Weinberg equilibrium, the frequency of the two alleles must add up to 1.0. Therefore, the frequency of the *D* allele must be 1.0 − 0.6, or equal to 0.4.

4. **The answer is D.**

If a population is in Hardy-Weinberg equilibrium, the frequency of *1,1* equals the frequency of *1* squared, or the square root of 0.49 equals 0.7. The frequency of *2* should be equal to 1 − 0.7, or 0.3. Then *2,2* is equal to 0.3 squared, or 0.09, which it is. The frequency of *1,2* is 2pq or 2 × (0.7 + 0.3), or 0.42. Thus, population *D* is in Hardy-Weinberg equilibrium. None of the other populations fit these predictions, and thus none are in Hardy-Weinberg equilibrium.

5–12. **The answers are: 5-A, 6-C, 7-B, 8-D, 9-A, 10-B, 11-C, 12-D.**

Pedigree **A** shows a typical autosomal recessive inheritance pattern with the following characteristics. On average, 25% of the offspring for a mating between two unaffected individuals are affected. Vertical transmission occurs, and affected individuals are not born in every generation. Consanguinity is present. Both males and females are affected.

Pedigree **B** shows a typical X-linked recessive inheritance pattern with the following characteristics. There is never father-to-son transmission of the affected allele on the X chromosome because the father transmits the Y chromosome only to his sons. Transmission of the X-linked affected allele from father to daughter is 100%, resulting in all daughters of an affected father being carriers. A heterozygous mother will transmit the trait to 50% of her sons, who will be affected, and to 50% of her daughters, who will be carriers. There is a much greater chance that males will be affected than females. Affected males are related to each other through carrier females.

Pedigree **C** shows a typical autosomal dominant inheritance pattern with the following characteristics. There is a 50% chance that any child of an affected individual will be affected. Affected individuals are seen in every generation. Unaffected individuals do not carry the defective allele and thus do not transmit the affected allele to their offspring. Both males and females are equally affected.

Pedigree **D** shows typical mitochondrial inheritance with the following characteristics: The disorder is maternally inherited. An affected father does not transmit the disorder to any of his children, whereas an affected mother transmits the disorder to all of her children.

PKU is an autosomal recessively inherited disorder, DMD is an X-linked recessive disorder, the Li-Fraumeni syndrome is an autosomal dominant disorder, and MERRF is a disease that has mitochondrial inheritance.

13–17. **The answers are: 13-D, 14-A, 15-G, 16-E, 17-E.**

Type II osteogenesis imperfecta is the lethal form of the disease in infants and presents with severe deformity and death within 1 month. A missense mutation occurs in the C-terminal end of the gene and results in an altered protein that interferes with normal assembly. The four common mutations listed occur in the β-glucosidase gene and are associated with Gaucher disease.

The fragile X syndrome is the most common inherited form of mental retardation known and is associated with an expansion of the trinucleotide CGG repeat in the noncoding 5′ end of the first exon of the *FMR1* gene.

The presence of café-au-lait spots and at least two Lisch nodules on the iris is diagnostic of NF-1. Additionally, benign neurofibromas, auxiliary freckling, and optic gliomas occur. The NF-1 gene codes for neurofibromin, a protein involved in the inactivation of Ras. Mutations in the neurofibromin gene are associated with NF-1.

18–22. **The answers are: 18-D, 19-B, 20-A, 21-C, 22-E.**

HNPCC, a common form of colon cancer, results from mutations in known human genes that function to keep mismatches from remaining in DNA. These genes code for proteins that are components of a DNA repair system known as the MMR system. Familial hypercholesterolemia, a result of a defect in the LDL receptor, is characterized by myocardial infarctions before the age of 55 years and increased cholesterol levels. This disorder is considered to be a haploinsufficiency disease, because half of the normal level of LDL receptor is insufficient to prevent clinical symptoms.

CF results from defects in the CF transmembrane conductance regulator. The most common mutation associated with CF is a specific 3-bp deletion in exon 10 (Δ F508) that leads to a loss of a phenylalanine residue at position 508 in the protein.

FAP is characterized by multiple benign colorectal tumors, some of which eventually develop into carcinomas. The *APC* gene codes for a protein that regulates the levels of β-catenin. In the absence of normal control of β-catenin, normal cell growth regulation is lost, increasing the probability of tumor development.

Myotonic dystrophy is one of the trinucleotide expansion disorders, but it is unusual in that the expansion region occurs in the 3′ untranslated region of the gene for cAMP-dependent kinase. The clinical symptoms of the disease appear to be related to a dominant effect of an RNA molecule that has multiple (up to 2000) copies of the CUG repeat at its 3′ end. This mutant RNA molecule sequesters CUG-binding proteins and disrupts normal processing of RNA transcripts in the nucleus.

PART III
Mapping and Cloning of Human Genes

41 Nucleic Acid Hybridization

A Nucleic acid hybridization

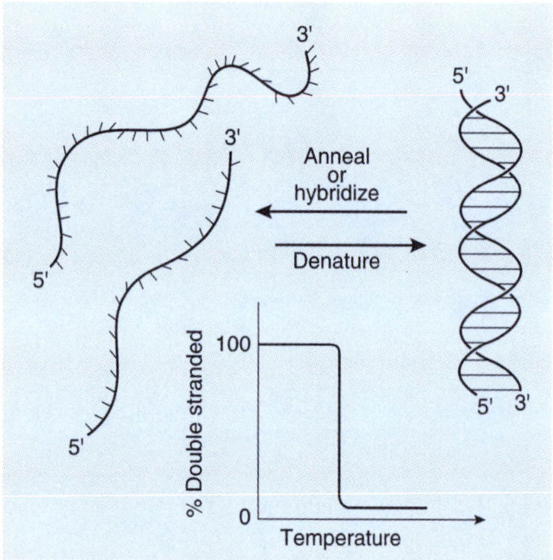

B Nucleic acid probes

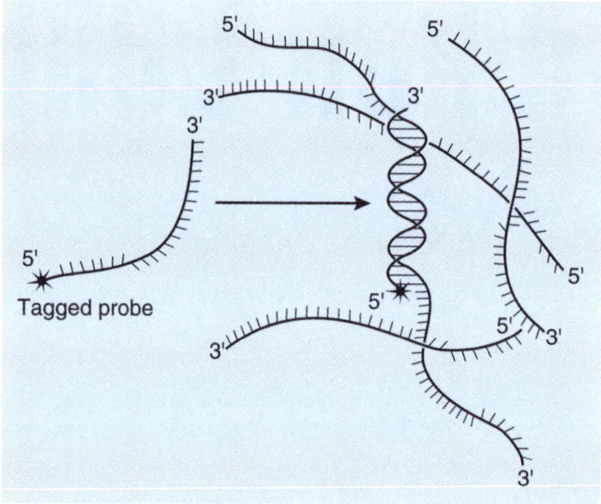

C Southern blot analysis

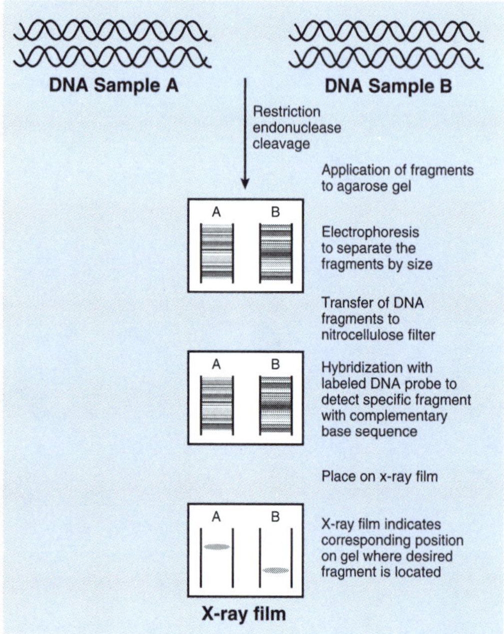

D DNA fragments detected on x-ray film

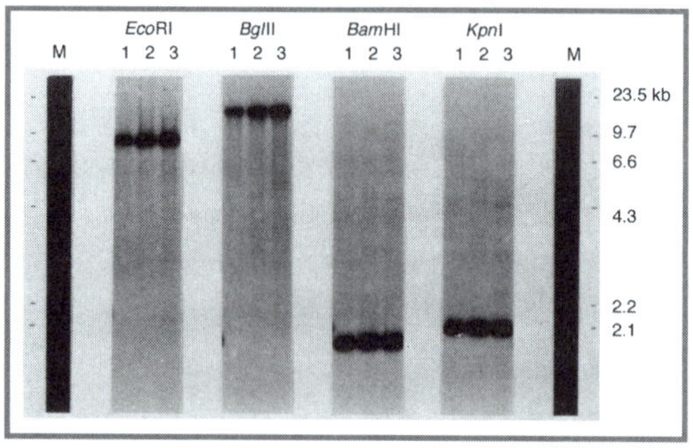

OVERVIEW

The principles of nucleic acid hybridization form the basis for many of the current technologies used to isolate and study normal and defective human genes. Nucleic acid hybridization occurs when two single strands of DNA or RNA (with complementary base sequences) are mixed together. Under the proper environmental conditions, the two strands will anneal to each other, or hybridize, to form a double-stranded structure.

Hybridizing Nucleic Acids

In the case of a DNA molecule, which is generally present as a double-stranded structure, it is necessary to treat or denature the DNA to separate the two strands, either by heating the solution of DNA or by treating it with alkaline. As the temperature of the solution increases, a double-stranded DNA molecule denatures to form single-stranded structures. Salt concentration, such as the concentration of sodium ions (Na^+), has the opposite effect and tends to stabilize the double-stranded helix. By controlling the temperature and the Na^+ concentration of a solution of DNA, it is possible to favor either the denaturation of a double-stranded DNA molecule to single-stranded molecules or the hybridization of two complementary single-stranded molecules to form a double-stranded helix, as shown in Figure **A**. When conditions are altered (e.g., by lowering the Na^+ concentration, increasing the temperature), mismatched bases tend to denature. Conditions of hybridization stringency can be selected that will cause instability in a duplex with one single mismatched base pair. This concept of stringency is very important in identifying a single base change in a mutant DNA, such as is needed for allele-specific oligonucleotide hybridization (ASOH) (see Chapter 43).

Nucleic Acid Probes

The key to using nucleic acid hybridization to study genes is labeling a particular molecule of DNA or RNA so it can be detected even if it is in the midst of 1 million unwanted molecules. This identification relies on the availability of nucleic acid probes that can be labeled and used to detect specific DNA fragments. Three major types of nucleic acid probes are used: conventional DNA probes, RNA probes, and oligonucleotide probes.

Conventional DNA probes are isolated by the cloning procedures described in Chapter 44, using plasmid or virus-based vectors. RNA probes can be made by transcribing a DNA sequence in the presence of regulatory sequences on the vector. Oligonucleotide probes are generally 30–50 nucleotides in length and are synthesized in vitro. To synthesize an oligonucleotide probe, either some of the DNA or RNA sequence must be known, or the known amino acid sequence of a protein can be used along with knowledge of the genetic code. When amino acid sequence information is used, it is necessary to use multiple oligonucleotides as the probe to compensate for the degeneracy of the genetic code. In all cases, the probe must be tagged or labeled so that after hybridization the resulting molecule can be detected.

Numerous methods are available to label nucleic acid probes. Radioisotopes such as P^{32} can be incorporated into the probe so that the probe and any nucleic acid molecules to which it hybridizes can be detected by autoradiography. (Autoradiography is performed by placing a sample on x-ray film. The decay of the isotope causes a dark area on the film, allowing the detection of the labeled nucleic acid.) Nonradioactive labeling systems—in particular, labeling probes with fluorescent compounds coupled to biotin labeling of the probe—are widely used. Nucleic acid probes are labeled with a biotinylated nucleotide such as dCTP and then used in the hybridization reaction. A protein called avidin, which comes from egg whites, has a very strong affinity for biotin and is added to the reaction mixture, wherein the avidin binds to the biotin-labeled probe. Then, an anti-avidin antibody containing a fluorescent label is added and binds the avidin. When the sample is placed under a fluorescent detector, it gives off a signal that locates the position of the probe and detects a specific DNA fragment. This technique is used in the FISH analysis discussed in Chapter 4 to locate DNA sequences at specific sites on a chromosome in a metaphase spread. In Figure **B**, the process of using a tagged probe is demonstrated. The tagged probe is mixed with a solution of unlabeled DNA molecules under conditions that favor hybridization of single strands with complementary base sequences. The probe specifically hybridizes to the DNA with the complementary sequence, and that DNA fragment becomes labeled, which allows it to be detected among the many millions of molecules that are not hybridized to the probe.

Southern and Northern Blot Hybridization

Southern blotting (Figure **C**) has been used extensively over the past 20 years to detect target DNA fragments after size fractionation by gel electrophoresis. A purified DNA sample containing the particular gene of interest, the target gene, is treated with a restriction endonuclease to produce many DNA fragments of varying sizes. The solution of fragments is applied to an agarose gel, and an electrical current is added, which separates the gel fragments according to size. The smaller fragments move faster and farther down the gel. At this point, the Southern blot is performed by denaturing the DNA on the gel in an alkaline solution, then transferring the denatured fragments from the gel to a nitrocellulose membrane filter by blotting. The denatured DNA fragments, which are immobilized on the membrane filter, are then allowed to hybridize with a labeled nucleic acid probe that can detect the target DNA. The filter is placed on an x-ray film to detect the site where the probe binds, as indicated by a darkening of the film (Figure **D**). The position on the membrane filter where the probe binds indicates the presence of the target DNA. When this position is compared with the original gel and molecular weight standards, it can give an estimate of the size of the target DNA fragment. A variation of the Southern blot technique is the Northern blot analysis, in which the target nucleic acid is an RNA molecule instead of a DNA molecule.

For more information see Hoffee PA: *IMS: Medical Molecular Genetics*. Fence Creek, pp. 114–119.

42 Methods for Detecting Defective Genes I

The Polymerase Chain Reaction

A Polymerase chain reaction

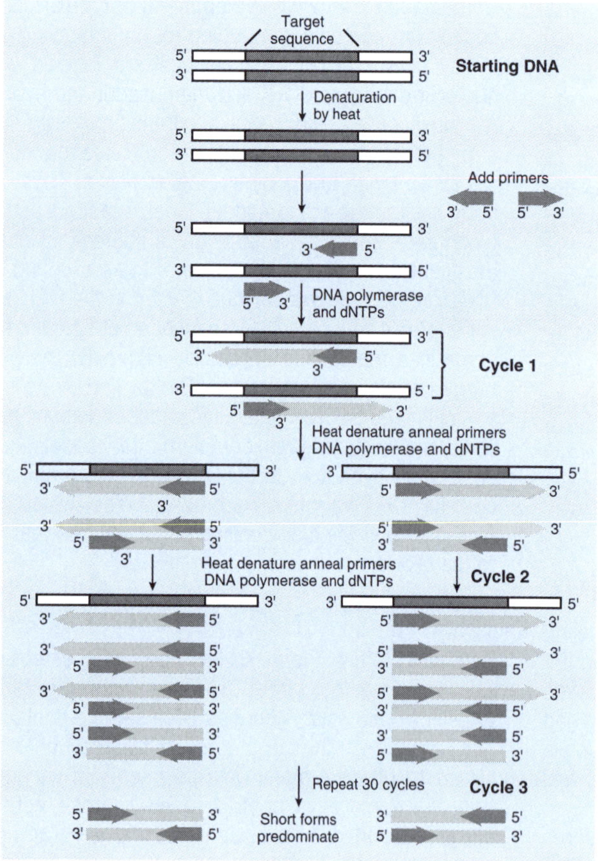

B Detecting PCR products by ethidium bromide

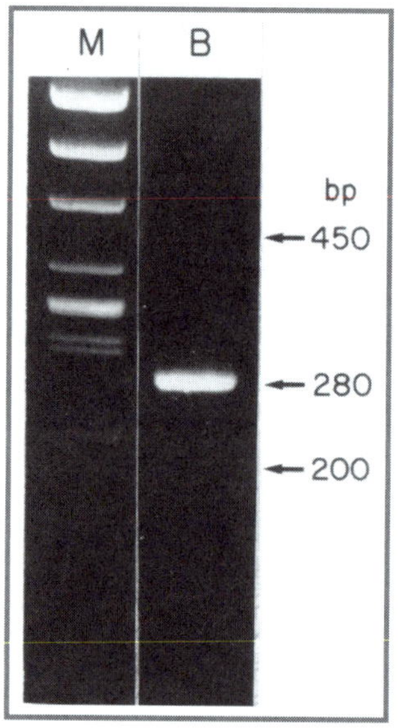

C Amplification of CA repeats by PCR

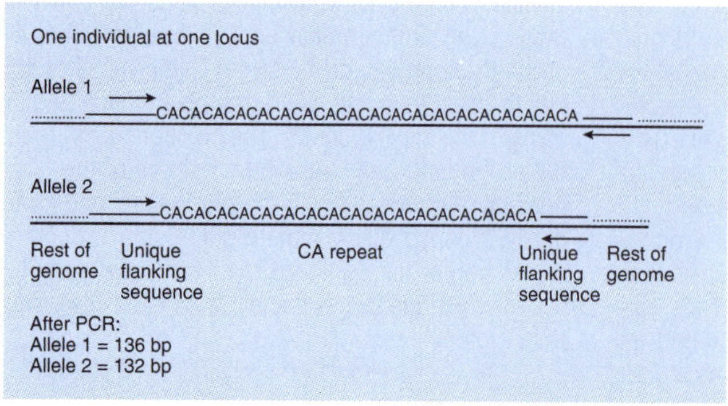

D Detecting PCR products in an automated DNA sequencer by fluorescence

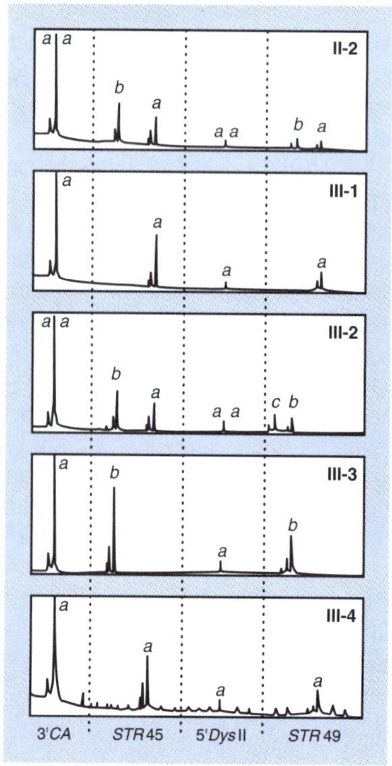

OVERVIEW

One of the most recent and innovative techniques developed for cloning and analyzing DNA is the polymerase chain reaction (PCR). With the use of PCR, it is now possible to make millions of copies of a specific DNA sequence in vitro in a few hours, and it can be done without first separating that DNA fragment from the rest of the genomic DNA. This extraordinary technique, developed by Kary Mullis in the mid-1980s, has revolutionized the way scientists analyze DNA in the detection and study of many human diseases.

Polymerase Chain Reaction

The PCR is diagramed in Figure **A**. To perform PCR, one begins with a double-stranded piece of DNA with the 5′ and 3′ ends, as noted in Figure **A**. Only a few molecules of starting DNA are necessary to begin the process, and they need not be purified away from other DNA molecules. In the first step of the reaction, the double-stranded DNA is denatured by heating to give two separated single strands. Added to the mixture of denatured DNA are two oligonucleotide primers (each 20–30 nucleotides long). One primer (the forward primer) is a 5′ to 3′ oligonucleotide that is complementary to one strand of the DNA; the second primer (the reverse primer) is a 5′ to 3′ oligonucleotide that is complementary to the other strand of the DNA. The two primers are added in excess so they will be present in sufficient concentration to carry out 30 cycles of the reaction. Each primer has a 3′ free end that serves as a substrate for DNA synthesis in the 5′ to 3′ direction, using each initial DNA strand as a template. After the oligonucleotide primers are annealed to each template DNA strand, DNA polymerase is added along with the precursor deoxyribonucleotides. DNA synthesis proceeds down each DNA strand, making a new copy of the DNA and giving now two double-stranded DNA molecules where there was initially only one. The 5′ ends of the newly synthesized strands are fixed, but the 3′ ends may vary in length. This ends the first cycle. At this point, the two double-stranded DNA strands are again denatured by heating, giving four single strands of DNA to serve as templates for new DNA synthesis. Once again, the primers bind to each single strand of DNA, and DNA polymerase begins synthesis at each primer, resulting in eight single strands of DNA to serve as templates in the next cycle. In the third cycle, the length of the newly synthesized DNA strands begins to become fixed, the 5′ ends defined by the primer and the 3′ ends defined by the terminus of the opposite primer. The cycle is repeated again and again, resulting in the number of DNA molecules doubling at each cycle with the short form that contains the target sequence beginning to predominate. After 30 cycles, nearly a billion copies of the DNA sequence located between the two primers is produced.

Detecting PCR Products

A billion copies of a DNA sequence is sufficient to allow detection directly on an agarose gel, either by fluorescence, if a fluorescent-labeled primer is used, or simply by staining with ethidium bromide. A product produced from PCR and its detection by ethidium bromide staining is shown in Figure **B**. Because the 30 cycles of the PCR reaction are carried out sequentially in a single reaction mixture, the DNA polymerase used in PCR, *Taq* polymerase, is isolated from a heat-resistant bacterium, *Thermus aquaticus*. The use of this heat-resistant polymerase allows the reaction mixture to be heated to 95° C for the denaturation step without inactivating the polymerase activity. It is important to remember that one requirement of PCR is prior knowledge of at least part of the base sequence of the DNA to make the necessary primers.

Detection of Short Tandem Repeat Polymorphisms by PCR

Short tandem repeats (STRs) are stretches of DNA sequence composed almost entirely of repeating cytosine–adenine dinucleotides. These stretches of CA repeats are found throughout the human genome and are often polymorphic, meaning that the precise numbers of repeats at any particular CA repeat locus varies. The exact number of repeats in a person's two alleles can be measured at any particular CA repeat locus using PCR amplification. In utilizing STRs to type alleles, primers are synthesized from unique base sequences that flank the STRs, allowing the amplification by PCR of alleles that differ by different numbers of repeat units, as shown in Figure **C**. In the example shown, allele 1 has a CA repeat of 136 bp, and allele 2 has a CA repeat of 132 bp, each allele representing one of the two homologous chromosomes. During synthesis by PCR, the PCR products are labeled with a fluorescent maker (e.g., fluorescein), which allows their detection in an automated DNA sequencer. The sequencer is an electrophoresis system with a fixed fluorescence detector that measures fluorescent bands of DNA as they migrate past the detector. The type of data obtained in the analysis of CA repeats in the dystrophin gene using PCR is indicated in Figure **D**. The automatic sequencer presents the data where the *x*-axis represents the time of electrophoresis (an indication of the increasing size of the PCR DNA product) and the *y*-axis is the fluorescent intensity of the PCR DNA product. The four repeat alleles (3′*CA*, *STR*45, 5′*Dys*II, and *STR*49) found within the dystrophin gene differ in size by a few base pairs, with each allele detected as a major peak on the electrophoretogram. The size of the allele to the left is the smallest, and that to the right is the largest. For each position there can be more than one allele noted as *a*, *b*, or *c*. In the example shown, the female marked II-2 has a single peak at the 3′*CA* position, marked *a*,*a*, which represents her two dystrophin genes at this position and indicates that she is homozygous at this locus. The use of CA repeats in linkage analysis is discussed in Chapter 48.

For more information see Hoffee PA: *IMS: Medical Molecular Genetics*. Fence Creek, pp. 114–119.

43 Methods for Detecting Defective Genes II

ASOH and SSCP Analysis

Genetics 43

A Detecting mutations using ASOH

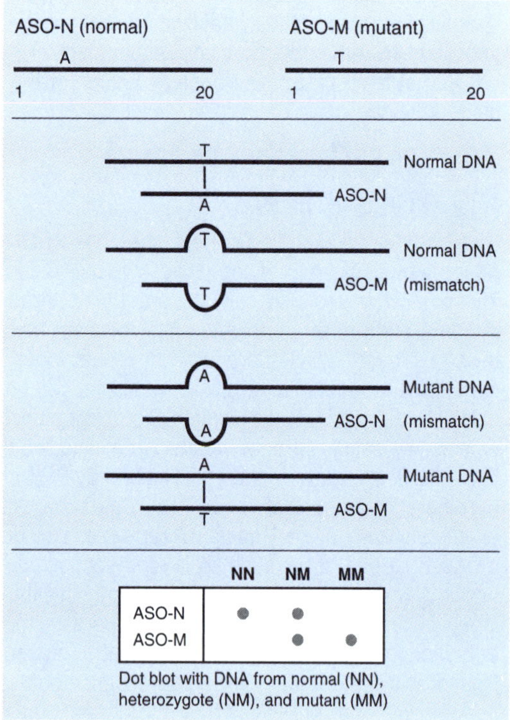

B Detecting mutations using SSCP analysis

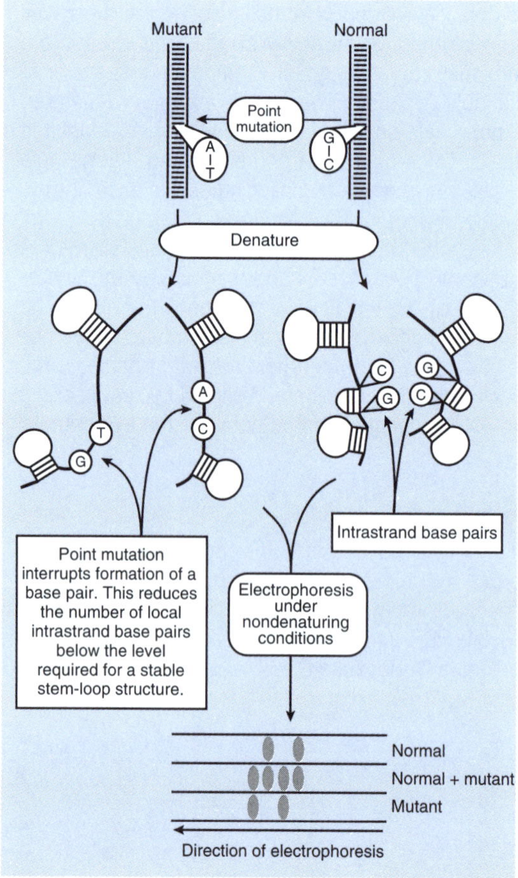

OVERVIEW

PCR has led to the development of two important techniques that can be used to detect single base change mutations in human genes. These two techniques, allele-specific oligonucleotide hybridization (ASOH) and single-strand conformational polymorphism (SSCP), are currently used in the clinic to analyze for mutations in many of the genes discussed in Chapters 27–37. These techniques can detect a heterozygous or carrier state as well as the homozygous state for a particular inherited disorder.

Detection of Mutations Using ASOH

Single base change mutations can be detected in genes if the sequence around the potential mutational site is known, for example, when dealing with mutations in the beta-globin gene (i.e., the mutation that results in sickle cell anemia). To detect these mutations, oligonucleotide primers that differ by a single nucleotide are synthesized in vitro and used with PCR-amplified DNA, as illustrated in Figure **A**. Two allele-specific oligonucleotides (ASOs) are synthesized; one is complementary to the normal allele (ASO-N) base sequence, and the other is complementary to the mutant allele (ASO-M) base sequence. The two oligonucleotides differ by a single base, A to T. The normal ASO hybridizes to the normal DNA allele but not to the mutant DNA allele, because a mismatch occurs. However, the mutant ASO hybridizes to the mutant DNA, but not to the normal DNA. When ASOH is used to detect mutations, DNA is isolated from a normal homozygote (NN), a heterozygote (NM), and a homozygous mutant (MM) and used as a control relative to unknown samples. The three control DNAs, along with the experimental samples, are spotted on a nitrocellulose membrane. Duplicate filters are individually treated with the radioactive ASOs. The normal ASO-N shows positive hybridization with the DNA obtained from the normal (NN) and the heterozygote (NM). The mutant ASO-M, however, shows positive hybridization to the heterozygote (NM) and the mutant (MM) DNA samples. Therefore, if the mutant oligonucleotide hybridizes to a DNA sample, it indicates that the patient from which that sample was taken is carrying the mutant DNA. Further analysis using the normal ASO distinguishes a heterozygote or carrier state from the homozygous mutant state. PCR is important in this assay because a very low amount of DNA (e.g., that present in a blood sample) can be tested by first using PCR primers to amplify the gene of interest and then testing the PCR-amplified DNA with the various ASO probes. When ASOH is used to detect mutations in DNA, the sequence change that has occurred to give the mutant phenotype and the surrounding sequences in the gene must be known so that PCR primers can be synthesized.

Detection of Mutations Using SSCP Analysis

SSCP analysis of PCR-amplified DNA depends on the tendency of single-stranded DNA to form intramolecular hydrogen bonding. The electrophoretic mobility of a DNA molecule in a nondenaturing gel system is a function of its conformation as well as its length. If a point mutation affects the intramolecular hydrogen bonding and stem-loop structure of the molecule, the DNA has an altered electrophoretic mobility. Thus, two single-stranded DNA molecules that differ by even a single base can be differentiated by SSCP analysis, as indicated in Figure **B**. A major advantage of SSCP analysis is that it can be performed when the mutation in the gene to be analyzed is unknown (unlike ASOH, which requires prior knowledge of the base change that leads to the mutation). When a mutation is suspected in a particular region of a gene, the DNA for the suspect mutant gene and the DNA of the normal gene are amplified by PCR. The amplified DNAs are denatured and immediately applied to a nondenaturing gel to prevent the denatured strands from renaturing. On the gel, the single-stranded DNA molecules remain separated from each other and begin to migrate in the gel. The native environment allows the single-stranded DNA molecules to hybridize with themselves and form intramolecular hydrogen bonding. This intrastrand hydrogen bonding occurs because the double-stranded form of DNA is more stable than the linear single-stranded form. The folding that occurs depends on the base sequence, with the two complementary strands folding to make different forms with an altered mobility in the gel. A single base change in 500 bases on either strand can be detected as an altered conformer or a mutant form of DNA. Further analysis of the altered DNA conformer can be carried out by isolating the DNA fragment from the gel and using automatic DNA sequence analysis to give the exact base change that has occurred. The base change identified in the DNA may or may not be related to the cause of the disease because any polymorphism that has occurred will give an altered DNA conformer. How do you determine whether the base change that has occurred is the cause of the disease?

Criteria to Establish That a Base Change within a Gene Is the Cause of the Disease

To determine whether the base change seen in SSCP analysis is the cause of the disease, several criteria must be fulfilled. First, one would expect the base change to result in an altered amino acid in the protein or the appearance of a stop codon in the mRNA. This prediction can be tested by examining the genetic code and determining whether the mutation seen leads to a change in a codon which would result in an altered amino acid or a termination codon. Second, if the disease is recessively inherited, one would expect to find both alleles or both copies of the gene to be altered in a patient with a mutant phenotype. For a dominantly inherited disease, only one of the alleles would be expected to be changed.

For more information see Hoffee PA: *IMS: Medical Molecular Genetics.* Fence Creek, pp. 119–123.

44 Molecular Cloning of Human Genes I

Restriction Endonucleases and Vectors

A Restriction endonucleases

Enzyme	*Recognition sequence*	*Bacterial source*
*Bam*HI	G↓GATCC	*Bacillus amyloliquefaciens*
*Eco*RI	G↓AATTC	*Escherichia coli RY 13*
*Hae*III	GG↓CC	*Haemophilus aegyptius*
*Hind*III	A↓AGCTT	*Haemophilus influenza Rd*
*Mst*II	CC↓TN*AGG	*Microcoleus* species
*Pst*I	CTGCA↓G	*Providencia stuartii*
*Sma*I	CCC↓GGG	*Serratia marcescens*
*Taq*I	T↓CGA	*Thermus aquaticus*

* N can be any nucleotide: adenine(A), thymine (T), cytosine (C), or guanine (G)

B Restriction endonucleases

*Pst*I (3' overhangs)	5' CTGCA↓G 3'	— CTGCA 3'
	3' G↑ACGTC 5'	3' ACGTC —
*Eco*RI (5' overhangs)	5' G↓AATTC 3'	5' AATTC —
	3' CTTAA↑G 5'	— CTTAA 5'
*Sma*I (blunt ends)	5' CCC↓GGG 3'	— CCC 3' 5' GGG —
	3' GGG↑CCC 5'	— GGG 5' 3' CCC —

C Constructing chimeric DNA molecules

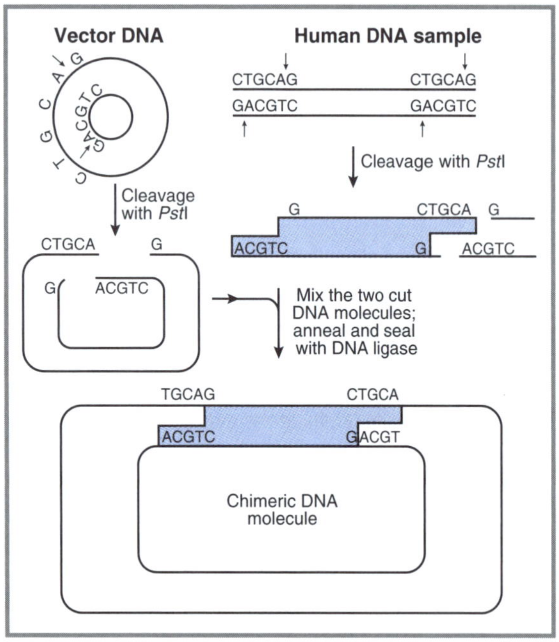

D Plasmid cloning vector

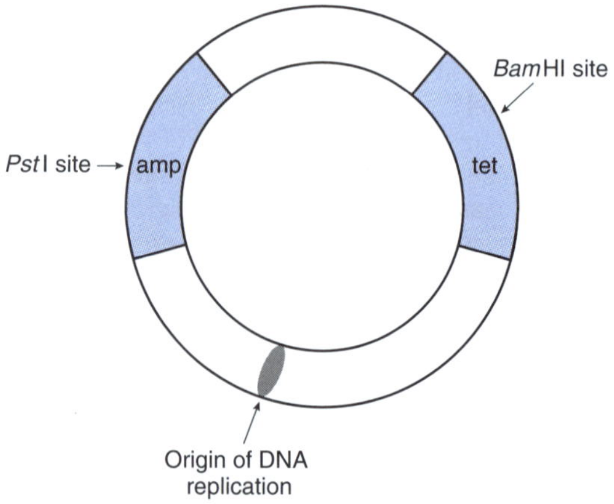

E YAC vector

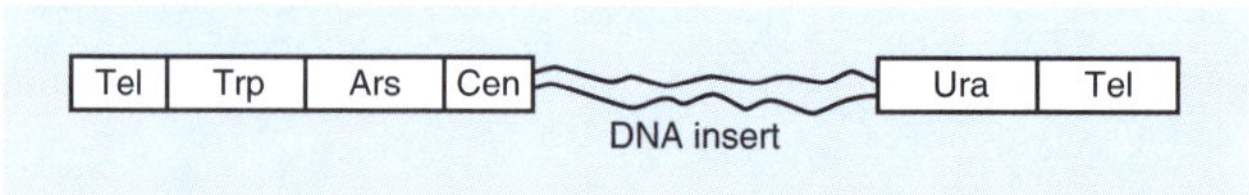

OVERVIEW

With increasing frequency, human genes are being isolated or cloned on a daily basis. Many of the techniques used in cloning human DNA fragments and genes were developed in the late 1960s and early 1970s with the study of bacteria and plasmids. These basic principles for isolating genes are applicable to the human genome even though it is a thousand times the size of a bacterial genome.

Restriction Endonucleases

The identification of bacterial enzymes, known as restriction endonucleases, that recognize and cleave short specific base sequences in DNA has been essential for the development of DNA cloning and has provided scientists with the ability to isolate and study genes from any species, including humans. A restriction endonuclease cleaves any double-stranded DNA molecule that contains the proper base sequence, whether the DNA comes from a bacterial cell, a yeast cell, or a human cell. This unique property allows a scientist to join together DNA fragments from any species to make a single chimeric molecule. The various restriction endonucleases, of which more than 100 are now known, are isolated from different bacterial species and are named accordingly; for example, *Eco*RI is from *Escherichia coli*, *Hae*III is from *Haemophilus aegyptius*, *Pst*I is from *Providencia stuartii*, and so on (Figure **A**). These enzymes, normally found in all bacteria, have evolved to protect the bacterial cells from infection by viruses. Viral DNA, when it enters a bacterial cell, is subject to degradation by the various restriction enzymes that prevent the viral DNA from replicating and killing the host cell. The bacterial DNA, although it also contains the sites for cleavage, is protected from degradation by its own enzymes by the presence of methyl groups at the site of cleavage. Each restriction enzyme can be characterized by the 4–8-bp sequence recognized to be cut, and the type of cleavage that occurs. The sequences recognized are generally palindromes, in which the sequence of bases on either strand of the DNA is the same when read in a 5′ to 3′ direction. In some cases, the DNA is cleaved to give blunt ends, but with most restriction enzymes, cleavage of the two DNA strands is offset by several bases, giving either a 5′ overhang or a 3′ overhang. An overhang is a short single-stranded end on the DNA molecule, sometimes referred to as a sticky end. These sticky ends have the ability to hybridize to other DNA molecules that have a complementary sticky end and are a crucial part in the cloning of DNA fragments. The specificities of some of the known restriction endonucleases and the types of fragments produced by restriction endonuclease cleavage of DNA are illustrated in Figure **B**.

Constructing Chimeric DNA Molecules

When two DNA molecules (e.g., one from a plasmid DNA and one from human DNA) are cut by the same restriction endonuclease, fragments are produced with single-stranded overhangs. These single-stranded overhangs on the two cut DNA molecules have complementary base sequences and can hybridize not only to themselves but to the other type of DNA fragment as well. In the presence of the enzyme DNA ligase, these base-paired sequences will be covalently ligated together to give a chimeric molecule—that is, a single DNA molecule made up of fragments of DNA from two different sources, as illustrated in Figure **C**.

Vectors: Replicating DNA Molecules Designed to Carry Foreign Genes

DNA molecules to which a target DNA is linked are referred to as vectors. Vectors have several key properties that make them important in cloning genes. They include (1) the presence of an origin of replication, (2) sites where restriction endonucleases cleave, and (3) the presence of a gene that allows the selection of specific cells that contain and are replicating the vector DNA. The use of vectors allows one to specifically pick out or clone a particular DNA fragment from a large number of DNA fragments. The particular fragment is then amplified to a high copy number so it can be isolated in sufficient quantities for analysis. Commonly used vectors include bacterial plasmids, the bacterial virus lambda, and yeast artificial chromosomes (YACs). The important properties of a typical plasmid cloning vector are shown in Figure **D**. Plasmids are small double-stranded, closed, circular DNA molecules that exist extrachromosomally in bacterial cells. These elements can replicate independently of the host chromosome, producing, in some cases, 50–100 copies of themselves in a single cell. Plasmids are also distributed to each daughter cell during cell division, and in the presence of selective pressure, a population of 1 billion bacterial cells, each of which carries 50 copies of the plasmid, can be produced. Plasmids are normal components of many bacterial cells, and often these DNA elements can confer on the host cell additional growth properties such as antibiotic resistance. Molecular biologists have been able to use the antibiotic-resistant property of plasmids in DNA cloning by creating vectors that contain genes that confer antibiotic resistance. The presence of an antibiotic resistance gene on the vector provides a method for selecting bacterial cells that carry the plasmid of interest (i.e., the vector carrying the target DNA) (see Chapter 45).

Yeast Artificial Chromosomes (YACs)

YACs are DNA elements that contain telomeres, a centromere, a yeast autonomous replication sequence (ARS), and human DNA pieces ranging in size from 100,000 to 1,000,000 bp (Figure **E**). The ARS enables the YAC to replicate in yeast cells, whereas the telomere sequences (Tel) give stability to the YAC, and the centromere (cen) serves to promote distribution of the YAC to each daughter cell at cell division. The construction of human YAC DNA libraries has simplified the analysis of the human genome by reducing about ten-fold the total number of clones that are required to contain the entire human genome sequence in the library.

For more information see Hoffee PA: *IMS: Medical Molecular Genetics*. Fence Creek, pp.??????

45 Molecular Cloning of Human Genes II

Cloning a Human DNA Fragment Using a Plasmid Vector

A Typical cloning procedure to produce a genomic library

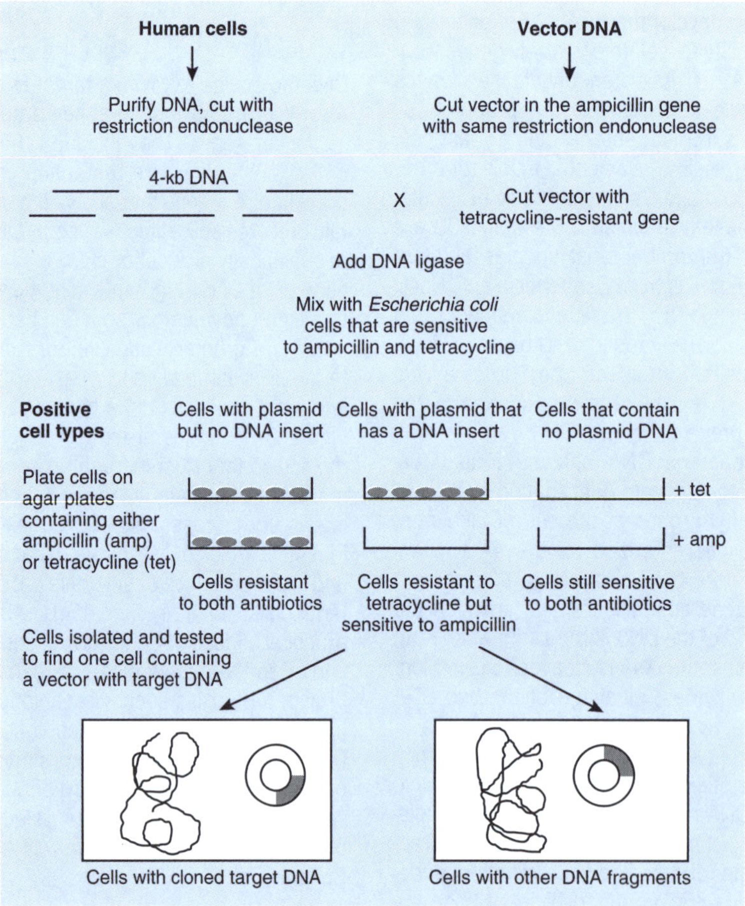

B Screening a DNA library

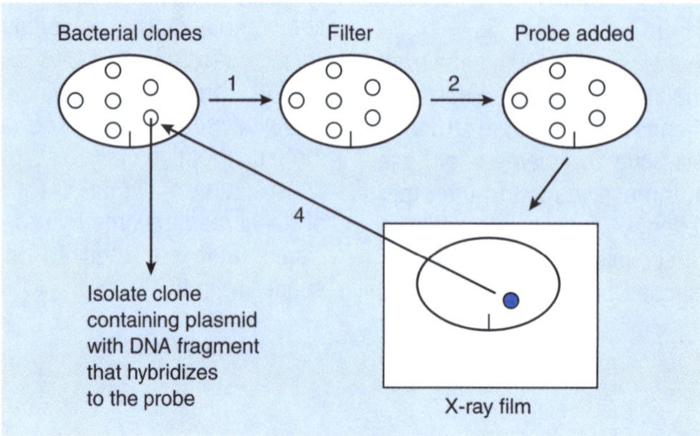

OVERVIEW

Each human cell contains 6 billion bp of DNA. However, the DNA fragment to be cloned (i.e., a 4000-bp fragment) represents only a minute fraction of the total. The first step in cloning a specific human DNA fragment is the purification of the DNA. This is followed by a series of steps to produce a collection of bacterial cells, each of which carries a vector with human DNA inserted into the *PstI* site. This collection of cells is referred to as a DNA library, or recombinant DNA library. Two major types of DNA libraries are constructed. When the DNA library is made starting with cDNA or DNA complementary to mRNA, it contains DNA sequences that represent only the coding regions of the genome. This is referred to as a cDNA library. When the DNA library is made starting with the total DNA isolated from cells, it contains DNA sequences that represent both coding and noncoding regions of the genome and is referred to as a genomic library. Each cell in a library made from human DNA should theoretically contain a unique fragment of human DNA; in a genomic library, the entire library should contain the complete sequence of human DNA genome.

Typical Cloning Procedure to Produce a Genomic Library

Purified DNA isolated from human cells is treated with a specific restriction endonuclease, such as *PstI*, which cuts the DNA at the sequence 5′CTGCAG3′ to give a 3′TGCA overhang (Figure **A**) (see Chapter 44). This digestion of the DNA by *PstI* produces thousands of fragments, only one of which is the one, the target fragment, to be cloned. A vector such as shown in Chapter 44, Figure **D** is used to isolate the target fragment. The vector DNA also is cut with *PstI* to give a linear molecule of DNA with single-stranded ends. The mixture of human DNA fragments containing the 4000-bp target DNA is mixed with the cut vector DNA in the presence of DNA ligase. The complementary single-stranded ends of the DNA molecules hybridize to each other and become covalently linked by the action of DNA ligase (see Chapter 44, Figure **C**). Three different types of recombinant DNA molecules result from this ligation: (1) a vector that has ligated back to itself and does not contain foreign DNA, (2) vectors that contain the various fragments of the human DNA produced by *PstI* digestion, and (3) a few vectors that contain the 4000-bp fragment to be cloned. With the use of antibiotic resistance markers, it is possible to eliminate any vectors that have not integrated a fragment of human DNA. This is accomplished by transforming an *E. coli* strain with the entire population of vector DNA and selecting for bacterial cells that have developed antibiotic resistance. The vector contains two antibiotic resistance genes, one conferring ampicillin resistance and one conferring tetracycline resistance (see Chapter 44, Figure **D**). The *PstI* site is in the middle of the gene that confers ampicillin resistance. If a vector has integrated a piece of foreign DNA at the *PstI* site, the gene conferring ampicillin resistance is inactivated. Therefore, transformed bacterial cells that carry a vector with a human piece of DNA inserted into the *PstI* site are sensitive to ampicillin but remain resistant to tetracycline. The transformed bacterial cells are of the following types: (1) cells that are resistant to both antibiotics and contain a vector without human DNA, (2) cells that are sensitive to ampicillin but resistant to tetracycline and carry a vector that has integrated a piece of human DNA at the *PstI* site, and (3) cells that are sensitive to both antibiotics and do not carry any vector DNA. Each of the ampicillin-sensitive, tetracycline-resistant cells (class 2) is then isolated and tested to determine which cell has a vector carrying the 4000-bp target fragment to be cloned.

Screening a DNA Library

The total collection of the cells carrying a vector with human DNA inserted into the *PstI* site (the DNA library) is screened to find the cells carrying the target DNA fragment. The normal human diploid cell contains 6 billion bp of DNA. The average size of a DNA fragment produced from a *PstI* digestion of human DNA is estimated to be about 7000 bp. Therefore, for a library to represent all of the *PstI* fragments, a minimum of about 1 million recombinant clones is necessary. However, only a few of these clones will carry the 4000-bp target fragment. (The term *clone* here refers to a population of genetically identical bacterial cells that have been derived from a single bacterial cell. For our purpose, the clone will be a population of bacterial cells, all of which contain a vector with the same DNA insert.) Two major approaches used to identify clones with specific DNA fragments involve looking either at the DNA sequence or at the protein or RNA that might be produced from that DNA sequence. To use the sequence approach, it is necessary to have a probe, which is a tagged RNA or DNA molecule that can be used to identify a clone carrying a particular DNA sequence using nucleic acid hybridization techniques (see Chapter 41). Each cell in the DNA library is grown on agar plates into a bacterial colony, and the clones are then transferred to a nitrocellulose filter (Figure **B**). The filter is treated to denature the DNA within the cells, and the radioactive probe is then added to the filter. If a cell contains a DNA fragment that is complementary to the probe, a radioactive spot is produced on the filter because of the hybridization of the radioactive probe to the DNA. These radioactive spots are detected by placing the filters on x-ray film. One then can go back to the original agar plate, pick the clone that hybridized to the probe, grow the cells to high density, and isolate the vector DNA in large quantities. If a nucleic acid probe is unavailable but the protein product that is produced by the DNA sequence is known, techniques are aimed at finding a clone that makes the particular protein or RNA of interest. However, this type of approach requires that the vector contain regulatory sequences that allow the transcription and translation of the target DNA, as well as the availability of an antibody made to the specific protein.

For more information see Hoffee PA: *IMS: Medical Molecular Genetics*. Fence Creek, pp. 110–114.

46 Physical Mapping I

Somatic Cell Hybrids

A Somatic cell hybridization

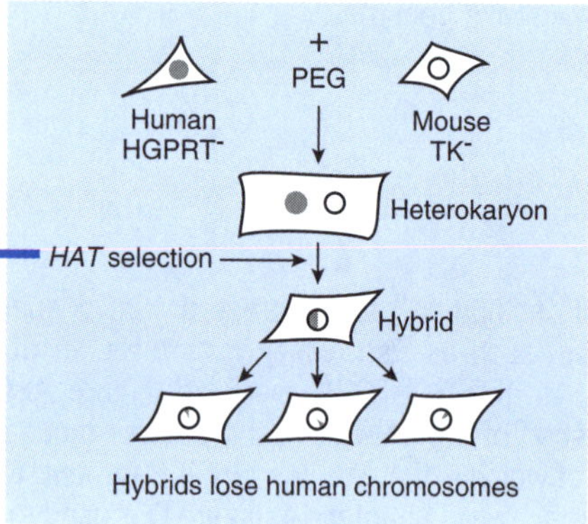

B HAT selection

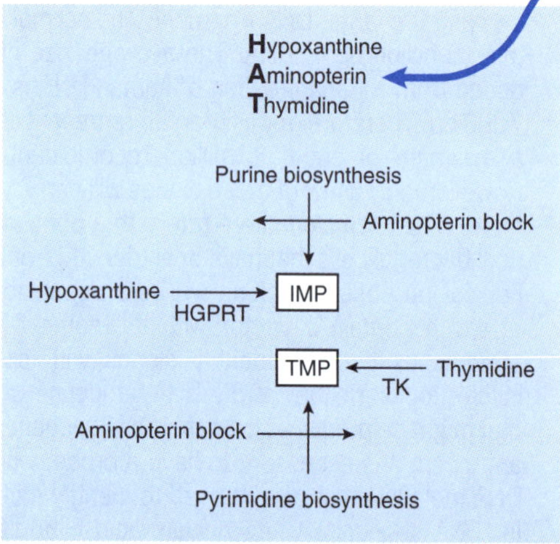

C Analysis of somatic cell hybrids

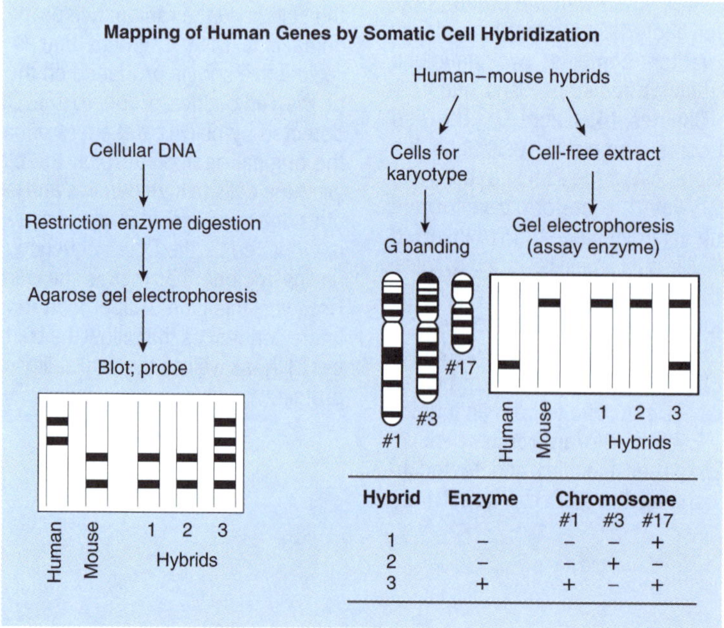

Hybrid	Enzyme	Chromosome #1	Chromosome #3	Chromosome #17
1	−	−	−	+
2	−	−	+	−
3	+	+	−	+

OVERVIEW

A physical map is the representation of the position of genes on each chromosome expressed in base pairs, with the genes placed in a known order and at a known distance from other genes. A physical map for the human genome is currently being constructed using results obtained from both cytogenetic and molecular analyses. The final goal in creating this physical map is the knowledge of the entire linear sequence of nucleotides for each human chromosome.

Somatic Cell Hybrids and Their Use in Human Gene Mapping

In the early 1970s, a method was developed for creating cells known as somatic cell hybrids, which have two entire mammalian genomes present in a single nucleus. In this technique, fusing agents (e.g., Sendai virus or polyethylene glycol) are added to cultured somatic cells, promoting the formation of multinucleated cells. When cells from two different species are fused together (Figure **A**), the initial heterokaryon contains the full chromosome complement from both species, but after nuclear fusion, the hybrid specifically loses chromosomes of one species and retains the chromosomes of the other species. Fusions of mouse or hamster cells with human cells generally result in the specific loss of human chromosomes and the retention of the rodent chromosomes in the hybrid nucleus. After 20–30 cell generations, a stable hybrid cell that contains the entire chromosome complement of the rodent, but a reduced number of human chromosomes, is formed. Because the loss of the human chromosomes occurs randomly, different hybrid cells in the population retain different human chromosomes. This unexpected random loss of human chromosomes from human–mouse hybrids was exploited as a new and innovative method for mapping human genes to specific human chromosomes. The assumption made in using somatic cell hybrids for mapping human genes is that a particular gene or gene sequence is always present in a hybrid when the chromosome on which it is located is present, and absent when the chromosome is absent from the hybrid. Therefore, a gene that is always associated with the presence of a particular chromosome is said to be syntenic with and localized to that chromosome. To isolate pure populations of human–mouse heterokaryons, a selection procedure must be used that kills both of the parental cells and the homokaryons but allows the human–mouse hybrids to survive and grow.

HAT Selection

The most common procedure used for the selection of somatic cell hybrids is HAT (hypoxanthine, aminopterin, thymidine) selection. In a medium that contains the drug, aminopterin, the de novo purine and pyrimidine biosynthetic pathways of cells are blocked. However, if hypoxanthine and thymidine are present in the medium, the cells can overcome the block by synthesizing their purines and pyrimidines using salvage pathways (Figure **B**). For cells to grow in HAT medium, two key enzymes, hypoxanthine-guanine phosphoribosyltransferase (HGPRT) and thymidine kinase (TK), must be functional. In the somatic cell hybridization illustrated in Figure **A**, the mouse cell is deficient in TK (TK^-) and the human cell is deficient in HGPRT ($HGPRT^-$). Neither mouse cells, human cells, nor homokaryons formed from these cells can grow in HAT medium because they are missing one of the enzymes needed for nucleotide synthesis. Heterokaryons, on the other hand, will have a normal *TK* gene that comes from the human genome and a normal *HGPRT* gene derived from the mouse genome. Therefore, in HAT medium, only human–mouse hybrids will survive and grow. As each isolated hybrid cell continues to grow and divide, daughter cells arise that have randomly lost human chromosomes, and eventually stable hybrid cells can be isolated, each of which has 5–11 human chromosomes remaining in the nucleus. However, each isolated stable hybrid will have a different set of human chromosomes remaining. In this way a hybrid cell panel is established, in which each hybrid retains only a few, but different, human chromosomes.

Analysis of Somatic Cell Hybrids

Several methods can be used to determine which human chromosomes are present in each hybrid clone (Figure **C**). Chromosome-banding procedures, such as G banding, were once used exclusively to identify the human chromosomes present in the hybrid. More recently, sets of chromosome-specific primers have become available that allow PCR screening to be used for identifying the human chromosomes that remain in the hybrids. Once the specific human chromosomes present in each hybrid have been identified, several approaches can be used to determine whether a particular human gene or gene sequence is present or absent from the hybrid. If the function of the particular gene is known, for example an enzymatic activity, or if a specific antibody to the protein coded for by that gene is available, the hybrids can be analyzed for the expression of the human protein. If the function of the gene is unknown but a DNA probe is available, then the presence of a specific gene sequence can be determined by Southern blot analysis carried out on DNA extracted from each hybrid. The presence of a specific human DNA fragment that hybridizes to the DNA probe indicates the presence of that gene in the hybrid, even though it may not be expressed. Once each hybrid in the panel has been analyzed with respect to the presence or absence of a particular chromosome and the presence and absence of a particular gene sequence, the assignment of that sequence to a specific chromosome can be made.

For more information see Hoffee PA: *IMS: Medical Molecular Genetics*. Fence Creek, pp. 136–141.

47 Physical Mapping II

Molecular Technology

A Isolation of individual chromosomes by flow cytometry

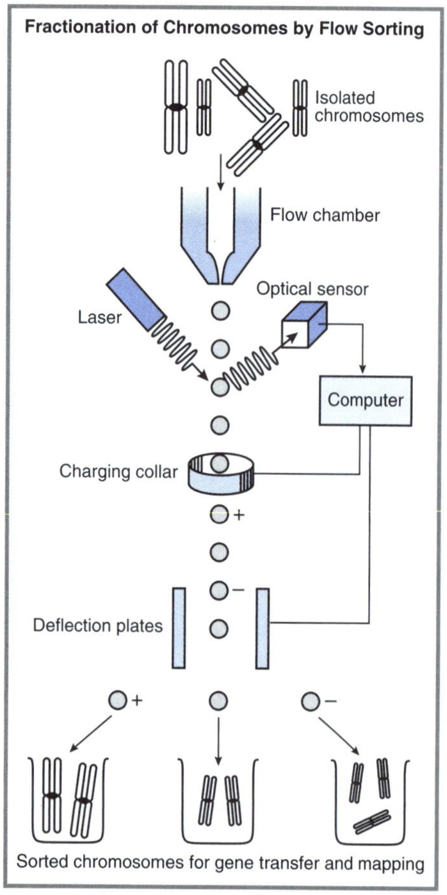

B Use of sorted chromosomes for mapping human genes

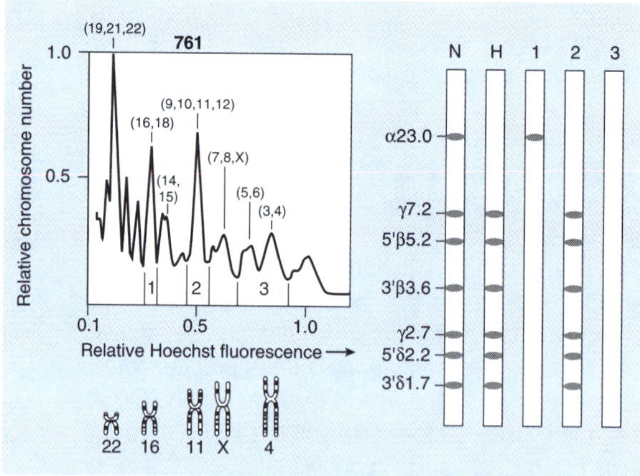

C Use of YACs and STSs in physical mapping

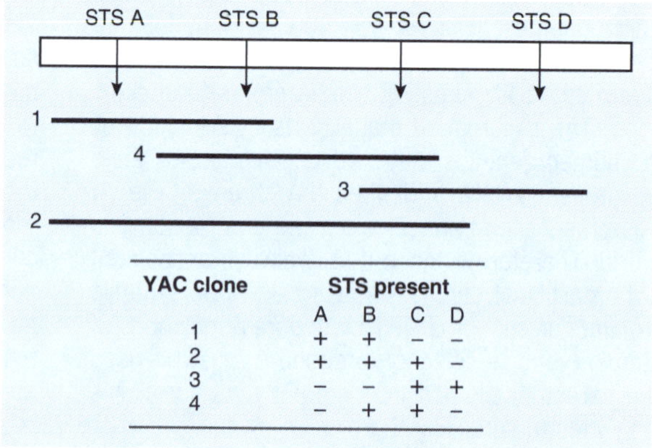

D Important techniques used in physical mapping

Approach	*Technique*	*Size resolution*
Chromosome	Flow cytometry, microcell hybrids	2×10^8 kb
Cytogenics	Banding, FISH	2×10^3 kb
YACs	Fragments cloned, overlapping order	100–1000 kb
BACs	Fragments cloned, overlapping order	150–350 kb
Restriction sites	Order restriction sites	200 kb
STS sites	Order STS sites on YAC clones	100 kb
Cosmids	YAC clones cut and recloned	40 kb
Plasmid/M13	Cosmid clones cut and recloned	1–5 kb
Sequencing	Each isolated fragment sequenced	1 bp

Note: *BACs* = bacterial artificial chromosomes; *FISH* = fluorescent in situ hybridization; *STS* = sequence-tagged sites; YACs = yeast artificial chromosomes

OVERVIEW

New molecular methods are now available to carry out the physical mapping of human genes. Many of these methods are currently being used in the Human Genome Project, which is designed to create a complete physical and genetic map of the human genome.

Isolation of Individual Chromosomes by Flow Cytometry

Individual human chromosomes for use in molecular studies can be isolated in pure form by the process of flow cytometry. In this process, a population of human chromosomes is isolated intact from cells and then labeled with a fluorescent DNA-binding dye that can be detected in a laser beam. Each chromosome absorbs a different amount of the dye depending on its size. Once labeled, the chromosomes are passed through a laser beam, and the intensity of the fluorescence is measured (Figure **A**). The chromosomes are then automatically sorted according to the intensity of the fluorescence and thus isolated according to size.

Use of Sorted Chromosomes for Mapping Human Genes

Because some human chromosomes are very similar in size, the separation in the flow cytometer is not always ideal, as is illustrated in Figure **B**. However, certain modifications of the procedure can be used to achieve a better separation of chromosomes of a similar size. Somatic cell hybrids that contain only a few human chromosomes can be used, simplifying the starting material and enabling a more precise separation of chromosomes of the same size. Particularly useful in this regard are human–Chinese hamster hybrids, because Chinese hamster DNA has a different base content than human DNA. This difference results in different dye binding and better separation of the rodent and human chromosomes. Additionally, chromosomes with translocations are valuable, since they usually differ in size and intensity of fluorescence from the normal chromosome. This difference in size results in an altered separation of the translocated chromosome in the flow cytometer and can be used to verify the position of a gene on a translocated chromosome. Once the human chromosomes have been isolated in pure form, they are used in Southern blot assays or dot blot assays for the direct localization of human genes or gene fragments to the specific chromosome (Figure **B**). An additional advantage of having purified individual chromosomes available is the ability to construct specific DNA libraries for each human chromosome. No longer is it necessary to begin with the total DNA content of a human cell in preparing a DNA library. Instead, the libraries can be made starting with DNA from each isolated chromosome, which reduces the number of clones that must be assayed. This is particularly true if the gene of interest has already been mapped by somatic cell hybridization or FISH to a particular chromosome. DNA libraries for each human chromosome are now available and are being used in conjunction with YACs and cosmids to determine the complete nucleotide sequence of the human genome.

Use of YACs and Sequence-Tagged Sites in Physical Mapping

The construction of human YAC DNA libraries has simplified the analysis of the human genome by reducing about ten-fold the total number of clones required to contain the entire human genome sequence in the library. Many of the YAC libraries have been constructed by partial digestion of isolated human chromosomes with a restriction enzyme to give DNA fragments that have overlapping sequences. The DNA fragments are separated with pulse gel electrophoresis to select for larger pieces of DNA and cloned into YAC vectors, which are propagated in yeast cells. Each isolated YAC clone will contain a specific segment of DNA of 100 to 1000 kb that must be analyzed and ordered with respect to every other DNA segment to form a composite map of each chromosome. The DNA fragments are analyzed for specific landmarks or sequences and then compared with each other in a search for overlapping regions on the different fragments. Ordering of the overlapping fragments into a continuous piece of DNA, a contig, allows for the construction of the final physical map. The use of sequence tagged sites (STSs) has greatly enhanced the physical mapping of human DNA and has provided a large number of landmarks on each chromosome. STSs are short fragments of DNA, about 200–500 nucleotides in length, which have a unique known sequence and thus can be amplified by PCR using forward and reverse primers with a specific oligonucleotide sequence. Once the positions and order of STSs are known on a chromosome, they can be used to order YAC clones. The presence or absence of various STSs on different YAC clones is determined by PCR, and these data are used to construct a STS content map. The assumption in putting together the map is that if the same STS is present on two YAC cloned DNA fragments, these fragments overlap. Overlapping fragments are then arranged to give a final physical map, as shown in Figure **C**. STSs can be known sequences derived from isolated genes or gene fragments or can be derived from CA repeat sequences used in linkage analysis. By combining the use of restriction endonuclease digestion sites, STSs, and YAC DNA libraries, a low-resolution physical map is now available for each human chromosome. A higher-resolution physical map is currently being constructed by cutting each DNA fragment in each YAC clone with a restriction endonuclease and subcloning these smaller DNA fragments into cosmid vectors, which contain inserts of about 40 kb. The cosmid DNA library is analyzed as above for landmarks to produce an overlapping sequence map now made up of smaller DNA segments. DNA fragments isolated from individual cosmid clones are then cut with a restriction endonuclease and subcloned into vectors that carry smaller fragments of DNA, such as plasmids or M13 phage. These smaller DNA fragments can then be purified and sequenced to give the complete DNA bp sequence for each fragment. Current mapping techniques are summarized in Figure **D**.

For more information see Hoffee PA: *IMS: Medical Molecular Genetics*. Fence Creek, pp. 141–145.

48 Genetic Mapping I

Linkage and Recombination Analysis

A Recombination frequency and genetic linkage

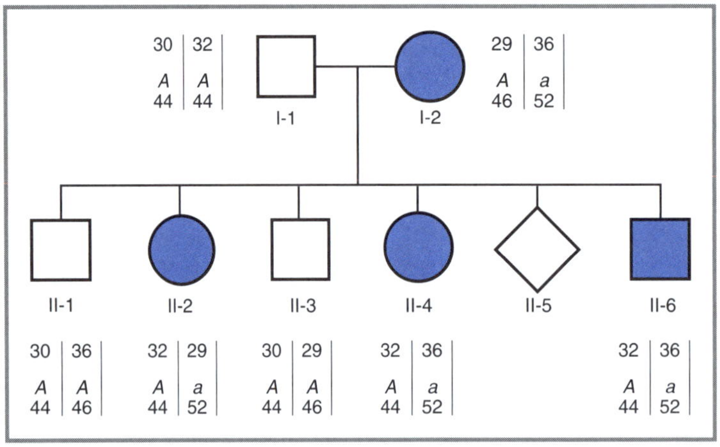

B Construction of genetic maps by linkage analysis

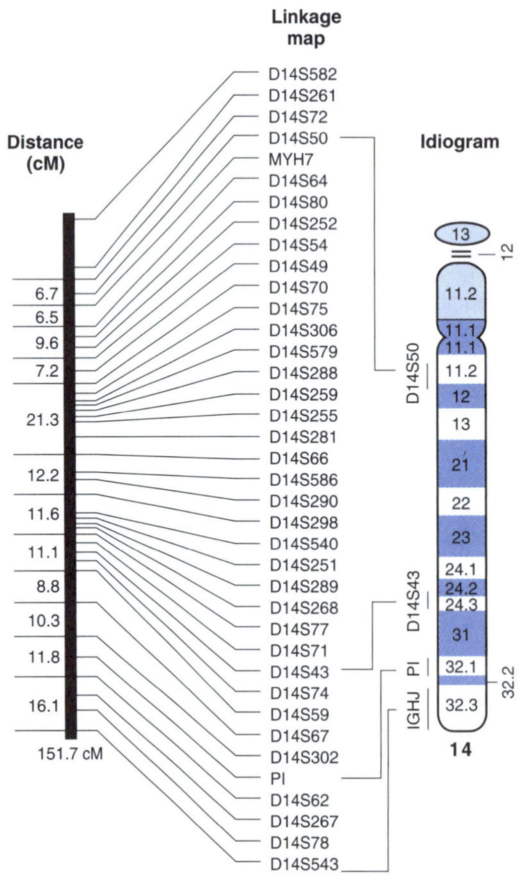

OVERVIEW

A genetic map is a representation of the relative position on a chromosome of a gene or gene locus based on the frequency of recombination between two segments of DNA during meiosis. This frequency is expressed in centiMorgans (cM); one centiMorgan equals a recombination frequency of 1%.

Recombination Frequency and Genetic Linkage

As noted in Chapter 8, during meiosis I, two homologous chromosomes can pair and undergo recombination, resulting in an equal exchange of genetic material between the two chromosomes, and producing recombinant chromosomes in the gametes. The probability that two gene sequences on a single chromosome will undergo recombination is dependent on the distance between the two sequences; that is, the closer the two sequences are together, the less probable it is that a recombinational event will occur between them, and the less probable it is that they will be separated from each other during meiosis I. As the distance between the two genes increases, the probability increases that a recombinational event will occur between them and that they will be separated from each other during meiosis I. Recombination frequency, therefore, is a relative measure of the distance between two loci on a chromosome, as shown in Figure **A**. In the figure, three positions are indicated on each of the two homologous chromosomes derived from the mother and from the father. The father is unaffected with the disease *a* and carries both normal alleles of gene *A*. The mother is affected with the disease and carries a normal allele, *A*, and a mutated allele, *a*. Two marker loci are indicated, one close to the disease gene and one further away. The marker closer to the disease gene does not show any recombination in the mother, because the children receive either *A*/46 or *a*/52. Because the mother is heterozygous and thus informative at the marker locus, an evaluation of linkage of the marker to the disease gene can be made. The father is homozygous for the close marker and the disease gene and is thus noninformative. The marker located farther away from the disease gene shows recombination in this small sample, indicating that crossing over has occurred between the marker and the disease gene. This high rate of recombination indicates that this marker is most likely not closely linked to the disease gene. Examination of the markers in the unborn child can be used to give a risk estimate of whether the child will be affected with the disease. If the unborn child receives the 52 marker from the mother, there is a 95% probability it will also receive the defective allele and have the disease. The measurement of recombination frequency, known as genetic linkage analysis, is based on the use of genetic markers, as shown in Figure **A**. Genetic markers are sequences in the DNA that are polymorphic, that are heterozygous in a random population, and that allow the detection of recombinational events occurring between two chromosomes. Several types of markers are used in genetic mapping. Originally, restriction fragment length polymorphisms (RFLPs) were used. These polymorphisms arise in the population when single base-pair mutations in the DNA alter the cleavage site for a restriction endonuclease. When the cleavage site is altered, cutting of the DNA with a restriction endonuclease will generate DNA fragments of different lengths, or RFLPs. The use of RFLPs in linkage analysis has been replaced for the most part by the use of variable number tandem repeats (VNTRs) and short tandem repeats (STRs). Tandem repeats are stretches of base sequences composed entirely of repeating di-, tri-, or tetranucleotides. They tend to be quite polymorphic in human DNA, and the precise number of repeats at any locus varies a great deal. In particular, CA repeats are used currently in linkage studies in combination with PCR analysis where PCR primers that flank the tandem repeat sequences can be used to detect and analyze each repeat sequence (Chapters 42 and 50).

Construction of Genetic Maps by Linkage Analysis

Linkage of genes on the X chromosome is often determined by linking a specific disease to X-linked inheritance and following the pattern of inheritance of the disease in a single family. Linkage of genes on autosomes, however, is much more difficult and depends on the availability of multigenerational families with a large number of children, living parents, and living grandparents. Unfortunately, for most inherited human disease genes, such families are not available, making it necessary that multiple smaller families be studied and the data analyzed statistically. These analyses are based on LOD score (logarithm of the odds ratio for linkage) calculations (Chapter 49). A comprehensive human linkage map consisting of 5840 loci, 970 of which have been ordered, has recently been constructed. This linkage map has markers every 0.7 cM and forms the basis of a high-density genetic map of the human genome. A typical linkage map for chromosome 14 is shown in Figure **B**. The map is created with the use of reference markers, short tandem repeats, and other polymorphisms. Markers, for example D14S74, are noted as D for DNA, 14 for chromosome 14, S for site of marker 74. Recombination rates generally are higher in females than in males; therefore, the distances in cM between markers on a human chromosome is greater for females than for males.

For more information see Hoffee PA: *IMS: Medical Molecular Genetics*. Fence Creek, pp. 145–149.

49 Genetic Mapping II

LOD Scores and Setting Phase

A Determining LOD scores

A LOD score greater than 0 is evidence in favor of linkage, whereas a LOD score less than 0 is evidence against linkage. By convention, a LOD score of 3.0 or greater (i.e., 1000:1 odds in favor of linkage) is *statistically significant* evidence of linkage, and a LOD score of -2.0 or less is statistically significant evidence against linkage. Note that the LOD score at $\theta = 0.5$ will always equal 0 (the $\log_{10}$ [1.0]).

B Setting phase

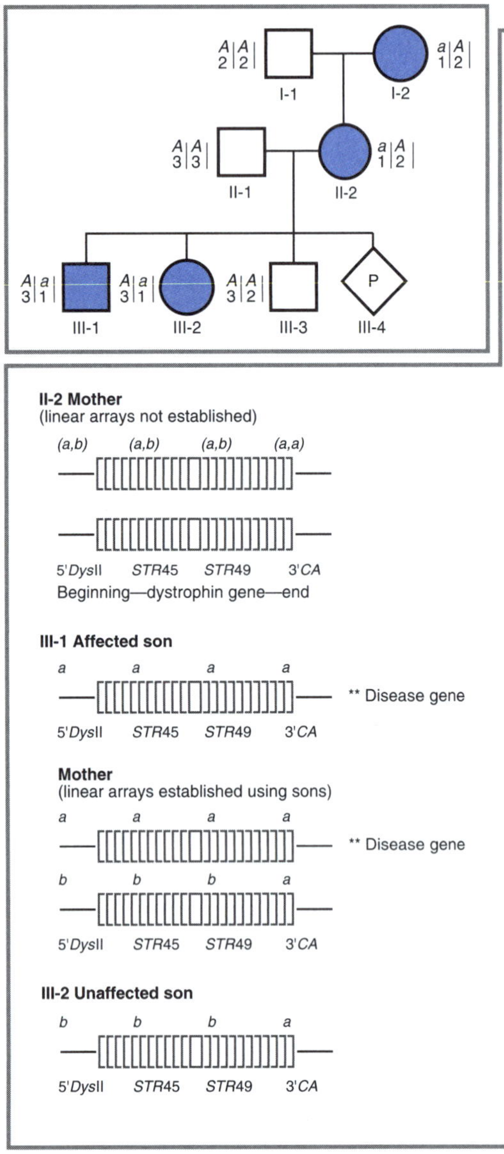

C Setting phase of X-linked disease genes with CA repeats

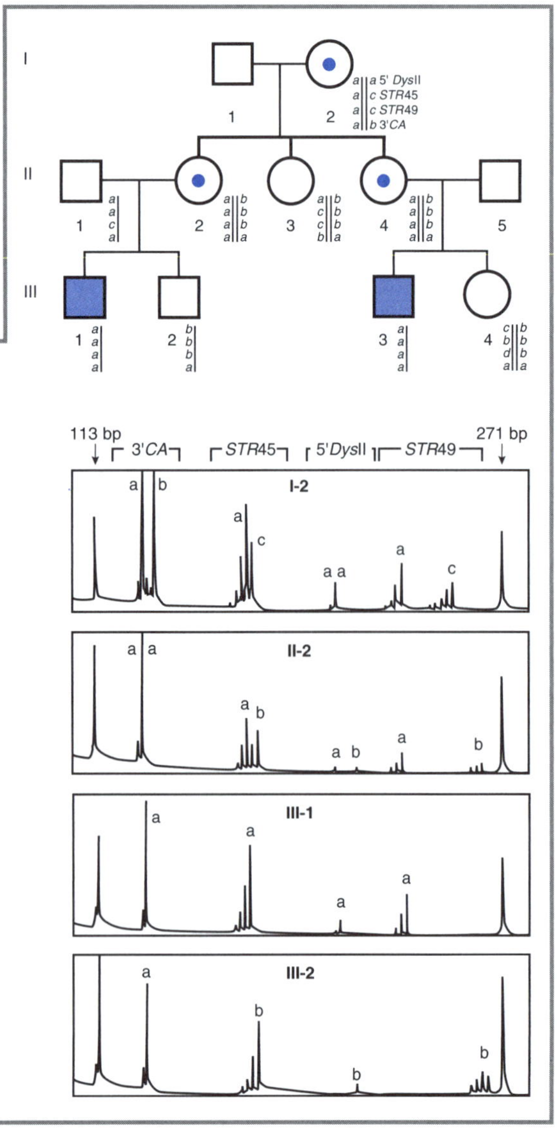

OVERVIEW

A determination of a LOD score is a statistical expression of confidence that a specific theta value between two markers has been identified (Figure **A**). Theta is defined as the distance between two regions of DNA expressed in units of recombination fraction. A 50% recombination between two regions is equivalent to a theta of 0.5 and is equal to no linkage. A 1% recombination between two loci is a theta of 0.01 and is equivalent to 1 cM, or about 1 million bp present between the two loci. Therefore, markers separated by more than 50 million bp (theta of 0.5) show no linkage.

Determining LOD Scores

An illustration of how LOD scores are determined is shown by examining 10 meiotic events in which no recombination between two markers has occurred, giving a theta of 0.00 (0% recombination). Statistically, the confidence of this theta is similar to the likelihood of flipping a coin 10 times and always getting tails, that is, 2^{10} (1:1024, or 1 in 10^3, or a log of odds of 3, LOD = 3). Thus, a LOD score of 3 or greater is considered significant that two genetic markers are closely linked. If a marker is found linked to a particular inherited human disease gene, it can be analyzed and followed through many families. How is linkage between a marker and a disease gene determined? Let us consider two loci: *A* and *B*. If 10% recombination is found between *A* and *B* in the families studied, that means 90% of the time there is no recombination between *A* and *B*. The recombination frequency, theta, is 10%, and the genes are roughly 10 cM apart. Are these two loci linked? To answer this question, one must calculate statistical support for the 10-cM distance. This calculation comes from the available data: the LOD score. As in the example above of flipping the coin 10 times, if we find 1 head and 9 tails in 10 flips (i.e., a recombination of 10%), the statistical support for an association between tails and flipping is weaker than if we flip tails 10 out of 10 times; the log of the odds (LOD) falls below 3. We can still make 10 cM statistically significant; however, we must flip the coin many more times and always observe the 9:1 ratio to obtain a LOD score of 3 or greater. In our study of markers *A* and *B*, many individuals, perhaps from many families, must be analyzed. If a LOD score of 3 or greater is found, we can say that the two markers *A* and *B* are in fact linked.

Setting Phase

To assess the recombination frequency and linkage between two loci, it is necessary to know phase information (i.e., is allele *A* on the same chromosome as allele *2* (in phase), or are the two alleles on different chromosomes (not in phase *2*). An illustration of knowing phase is diagramed in Figure **B**. In this pedigree, phase can be determined because the grandparents are available for analysis. The maternal grandmother, I-2, has the disease gene *a* and passed it on to her daughter, II-2, along with the marker *1* allele. The mother's *A2* combination comes from the father, who is homozygous at both alleles. Phase in the mother, II-2, is then *a1* on one chromosome and *A2* on the other chromosome. Of the four children, two are affected with the disease, one is unaffected, and one's genotype is unknown. Both children who received the *a1* haplotype are affected, whereas the child receiving the chromosome with *A2* is normal. No recombination events between the marker and the disease gene are seen in this small sample, consistent with linkage of the two loci. When phase is known, the probability that the unborn child will have the disease can be predicted. If the child receives marker *1*, the chances are that the child will also receive the disease gene and be affected. However, if the child receives marker *2*, the chances are that the child will receive the *A* allele and not be affected unless a recombinational event occurs between the two chromosomes.

Setting Phase of X-Linked Genes

For X-linked disease genes, phase can be set by the haplotype of an affected son. Because the single X chromosome in the son always comes from the mother, phase can be automatically set in the mother by determining the phase in the affected son (Figure **C**). The gene defective in Duchenne muscular dystrophy (DMD) is located on the X chromosome and contains four CA repeat alleles (5′*DysII*, *STR45*, *STR49*, and 3′*CA*). The different alleles (i.e., *a, b, c*, or *d*) are determined by use of PCR and an automated DNA sequencer, as indicated in Figure **C** (see Chapter 42). For the mother, II-2, at *STR45* there are two peaks marked *a* and *b*, indicating that her alleles differ at this locus. The 5′*DysII* and the *STR49* locus each show two peaks, *a* and *b*. Thus, this woman has an allele array as follows:

a,b	*a,b*	*a,b*	*a,a*
5′*DysII*	*STR45*	*STR49*	3′*CA*

However, without information from other members of the pedigree, the linear array of the markers on each of the two chromosomes cannot be determined. To determine this, we must "set phase." For example, in female II-2, is the order *b,b,a,a* on one chromosome and *a,a,b,a* on the other chromosome, or is some other combination present? To "set phase" with an X-linked disease, one starts with the males in the pedigree. Phase is already known in males, because males have only X chromosome and hence only one dystrophin gene. A schematic of the two dystrophin genes of the mother (II-2) is shown in Figure **C** along with the haplotypes of both of her sons (III-1 and III-2). The affected son, III-1, has an *a* allele at each locus, giving a pattern of *a a a a*. The unaffected son, III-2, has a pattern of *b b b a*. Each son must have received one X chromosome from the mother. Because one son is affected and one unaffected, the two sons probably received different X chromosomes (thus different dystrophin genes) from the mother. Therefore, the mother is most likely *a a a a* on one chromosome and *b b b a* on the other. We further can conclude that the "disease gene" has an *a a a a* pattern. Any female inheriting this gene is a carrier, and any male inheriting this gene (such as III-3) will be affected.

For more information see Hoffee PA: *IMS: Medical Molecular Genetics*. Fence Creek, pp. 107–108, 146–148.

50 Approaches to Cloning Human Genes

Functional Cloning, Positional Cloning, and the Positional Candidate Gene Approach

A Functional cloning

This cloning approach requires knowledge about the function and structure of the protein product of a gene but requires no prior information about the map position of a gene.

B Positional cloning

This cloning approach requires knowing the map position of a gene, but no prior information about the function or structure of the gene product is necessary.

C Identification of a disease gene by positional cloning

Goal	Steps
Localize gene to a specific site	Linkage analysis using multiple families Chromosomal abnormalities: Translocations, Deletions
Identify exon (expressing) sequences	Isolation of genomic DNA clones: Enlarge DNA size by chromosome walking Ensure that clone is correct (analyze patient DNA) Analyze DNA clones for the presence of exons: Carry out a zoo blot Perform Northern blot analysis on mRNA samples Screen a cDNA library Find an open reading frame by sequencing the cDNA
Identify function of gene	Use the cDNA base-pair sequence to determine an amino acid sequence of the protein Explore data banks to find homology to a known protein Use the cDNA in an expression vector to express the protein Make antibodies to the expressed protein Use the antibody to test normal and patient samples
Correlate the gene with the disease	Correlate mutational defects in the patient's genes with the disease phenotype and inheritance pattern Create a "knockout" mouse and reproduce the disease phenotype Correct the defect in vitro with a normal, functional gene

E Human disease genes isolated by positional cloning or the positional candidate gene approach

Disease	Location	Aberrant Protein
Alzheimer 1	21q	β-Amyloid precursor protein
Alzheimer 2	14	S182 membrane protein
Alzheimer 3	1q31–42	STM2 membrane protein
Ataxia telangiectasia	11q22–23	Phosphatidylinositol 3-kinase
Bloom syndrome	15q26.1	DNA helicase
Duchenne muscular dystrophy	Xp21	Dystrophin
Early-onset breast cancer 1	17q21	BRCA1 (nuclear protein)
Early-onset breast cancer 2	13q12–13	BRCA2
Cystic fibrosis	7q31.2	CFTR
Familial melanoma	9p21	p16
Fragile X syndrome	Xp27.3	FMR1
Hereditary nonpolyposis colorectal cancer (HNPCC)	2p15–22	hMSH2 (Mut S)
	3p21.3	hMLH1 (Mut L)
	2q31–33	hPMS1 (Mut L)
	7p22	hPMS2 (Mut L)
Huntington disease	4p16.3	Huntingtin
Li-Fraumeni syndrome	17p13	p53
Marfan syndrome	15q21.1	Fibrillin
Myotonic dystrophy	19q13	Protein kinase
Neurofibromatosis type 1	17q11.2	Neurofibromin
Neurofibromatosis type 2	22q12	Schwannomin
Retinoblastoma	13q14	Rb protein
Werner syndrome	8p12	DNA helicase

D Positional candidate gene cloning

This cloning approach relies on determining the map position of a gene and then a molecular analysis of the region for a predicted candidate gene.

OVERVIEW

One of the major goals of medical molecular genetics is the identification and cloning of all inherited human disease genes. The attainment of this goal began with the advent of recombinant DNA technology and the utilization of a strategy called functional cloning, which relies on knowing the biochemical structure of the protein involved, thereby using the protein as a means of isolating the corresponding gene.

Functional Cloning

In some inherited disorders, the basic protein defect in the disease is known at the biochemical level, and this enables the identification of the corresponding gene (Figure **A**). In these cases, the genes are isolated directly by recombinant DNA technology with no information about the map position of the gene. In functional cloning, the protein product of the gene must be known, purified, and usually sequenced. Knowledge about the amino acid sequence of the protein is used to make in vitro synthetic oligonucleotides that serve as probes to screen cDNA libraries. When a cDNA clone that is complementary to the oligonucleotide probe is isolated, it is used to screen genomic libraries in an attempt to isolate the complete gene sequence. Alternatively, antibodies directed against the protein can be used to look for bacterial clones in an expression library that produce the specific human protein. Many of the genes discussed in Chapters 27–32, which are associated with defects in metabolic pathways, were isolated by functional cloning.

Positional Cloning

As more genes were localized to specific human chromosomes by somatic cell hybridization, linkage analysis, and the visualization of chromosomal abnormalities in the light microscope, a technique known as positional cloning was developed for identifying and isolating human disease genes (Figure **B**). Positional cloning requires knowing only the chromosomal map position of a gene. No biochemical or functional information about the gene product is necessary.

Identification of a Disease Gene by Positional Cloning

With positional cloning, the location of a gene on a chromosome is determined either by pedigree and linkage analysis (Chapter 48) or by the identification of chromosome abnormalities (Chapter 55). A very important type of chromosome abnormality used in positional cloning is a translocation, which breaks the chromosome within a gene, disrupting the gene's function. If an inherited human disease is consistently associated with a common breakpoint in a chromosome, that point of breakage becomes an important clue to the position of the gene on the chromosome. Such a translocation was crucial in the identification and cloning of the gene for DMD (Chapter 51). Microdeletions that are consistently associated with an inherited human disease are also important clues to the position of a gene. In Prader-Willi syndrome, a microdeletion at 15q11–13 of the paternal chromosome is consistently associated with the disease. By use of this information, the gene for the snRNA associated polypeptide *SNRPN* was mapped to this position and shown to be associated with the disease. If no chromosomal abnormalities are found associated with the disease of interest, linkage analysis data must be used to localize the gene to a specific map position. Linkage information can resolve the map position usually to only within 1–3 cM or 1–3 Mb of DNA. Thus, additional information is needed before the desired gene can be definitely identified and isolated. The experimental techniques listed in Figure **C** used to identify the candidate gene in a large genomic fragment of 1–3 million bp are not simple and can be very time-consuming. Fortunately, more information is becoming available on the sequence and order of YAC clones and STSs containing DNA fragments from each human chromosome. As the base pair sequence of these YAC clones become known, the ability to identify and isolate a candidate gene from a particular position within a chromosome will become more feasible.

Positional Candidate Gene Approach

Genes corresponding to many proteins have been localized within the genome, although many of these genes have no specific human disease associated with them. The localization of genes has increased rapidly over the past few years because of the advent of expressed sequence tags (ESTs). ESTs are small cDNA sequences generated from mRNA molecules that have been isolated from specific human tissues. Currently there are 450,000 human sequences in the available EST collection, and they may represent nearly 80% of the total human genes. With the use of ESTs, many or all genes can be mapped to a specific region on a human chromosome. When a family segregating a disease gene is studied by linkage analysis and a locus is identified, then it can be determined whether a previously mapped gene (EST) is located in the same region of the chromosome. In this way, the gene identified by linkage analysis can be mapped to the EST site. This technique has been successful in several cases and is predicted to become the major strategy for the isolation of inherited human disease genes. The success of the positional candidate gene approach is directly related to the successful creation of a high-density EST map for the human genome (Figure **D**). Within the past few years, several genes have been isolated by this approach, including the genes implicated in hereditary nonpolyposis colon cancer (HNPCC) and the gene implicated in Bloom syndrome. Some examples of inherited disease genes isolated by positional cloning and the positional candidate gene approach are shown in Figure **E**.

For more information see Hoffee PA: *IMS: Medical Molecular Genetics*. Fence Creek, pp. 149–158.

51 Positional Cloning of the Duchenne Muscular Dystrophy Gene

A Inheritance pattern of X-linked DMD

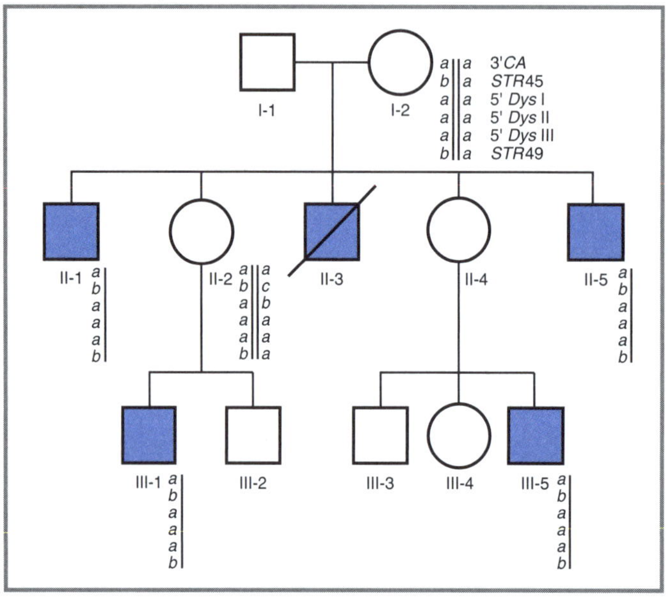

B X-autosome translocations associated with DMD

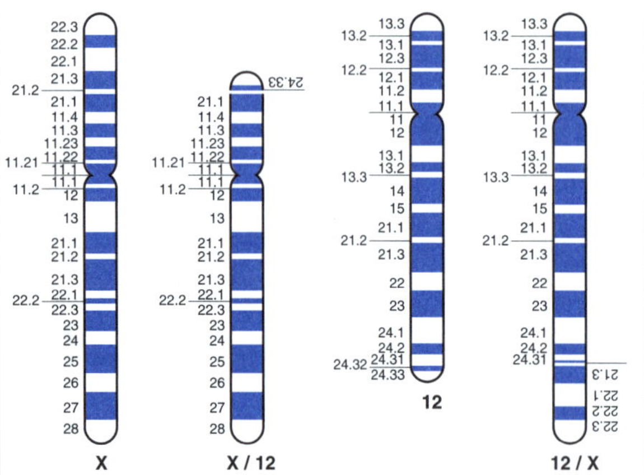

X Inactivation

Inactive X / 12

The cell is *not viable.* The cell has too many X chromosomes. One complete X chromosome is active, and part of the other X chromosome, which is on 12 / X, is also active. The cell does not have two copies of chromosome 12. Part of one chromosome 12 is on X / 12 and is inactivated.

Inactive normal X

The cell is *viable.* The cell has only one active X chromosome: part on 12 / X and part on X / 12. The cell has two normal active chromosomes 12: one normal and the other active on X / 12 and 12 / X.

Viable cells have an inactivated normal X chromosome.

It was hypothesized that each of the girls with an X-autosome translocation had DMD, with the function of the gene responsible for DMD being destroyed by a break at Xp21 that cut within the gene. However, only one of the DMD genes should have been affected by the translocation, leaving the other DMD gene intact. Thus, these girls should not show symptoms and should resemble nonsymptomatic carriers.

Why then were the girls affected with the disease? The answer lies in the process known as X inactivation, discussed in Chapter 57. In a human female, only one of the X chromosomes is functionally active; the other X chromosome undergoes inactivation, and most of the genes on the inactive X chromosome are not transcribed. Normally, this process is random. In some cells, one X chromosome is inactivated, and in other cells, the other X chromosome is inactivated. A problem occurs when part of the X chromosome is translocated to an autosome. X inactivation shuts down one X chromosome as usual. However, if the X chromosome with the translocation is the one shut down, the autosome piece also shuts down, and the cell dies. A cell survives if the normal X chromosome is preferentially inactivated. In that case, the cell has only one complete X chromosome active, both copies of the autosome are active, and the cell is viable.

If the normal X chromosome is preferentially inactivated, the normal DMD gene located on the normal X chromosome is also inactivated. Therefore, if the breakpoint at Xp21 occurs in the middle of the DMD gene, the female will have both DMD genes inactivated (one by X inactivation and the other by a disruption occurring in the middle of the DMD gene), and she will be affected by the disease.

OVERVIEW

The first example of positional cloning for identifying and cloning an inherited human disease gene was the isolation in 1987 of the DMD gene (Chapter 36). Before 1987, there was little or no information about the molecular basis for the muscle wasting in DMD, and nothing was known about the protein that was defective in DMD. This paucity of knowledge made the isolation of the gene by functional cloning highly improbable. The alternative approach was to use gene mapping information and to look directly for the gene without knowing anything about the properties of the gene or the gene product (i.e., positional cloning). If the gene could be isolated, then the function of its encoded protein could be determined.

Localization of the DMD Gene to a Specific Site in the Genome

The isolation of the DMD gene was simplified by knowing that the disease was X-linked and inherited as an X-linked recessive trait (Figure **A**). Location of the DMD gene on the X chromosome immediately reduced the total amount of DNA to be analyzed to 10% of the genome, but this amount of DNA still represents about 3×10^8 bp. A normal gene on the average represents only 3×10^4 bp, or less than 0.01% of the total DNA of the X chromosome. Next, it was necessary to localize the DMD gene to within a smaller region of the X chromosome.

X-Autosome Translocations Associated with DMD

An important piece of information that led to the positional cloning of the DMD gene was the observation of a series of young girls who had a disease that looked like DMD clinically and were shown by cytogenetic analysis to have an X-autosome translocation. Analysis of their karyotypes showed the presence of one normal X chromosome; however, part of their other X chromosome was translocated to an autosome (Figure **B**). All of the girls showed the breakpoint in the X chromosome in the same place (Xp21), but the autosome involved was different for each patient. It was hypothesized that each of these girls with an X-autosome translocation had DMD, with the function of the gene responsible for DMD being destroyed by a break at Xp21 that cut within the gene.

Additional Steps in the Isolation of the DMD Gene (see also Chapter 50, Figure C)

The isolation of the DMD gene was accomplished in the laboratory of Louis Kunkel with the use of subtraction cloning, a technique that looks for a piece of DNA that is present in the genome of a normal individual but is missing in a patient's deleted DNA. Dr. Kunkel used DNA from a boy who was affected not only with DMD but with several other diseases, including retinitis pigmentosa and chronic granulomatous disease. Analysis of the boy's DNA by cytogenetic chromosome banding had revealed a microdeletion of about 10 million bp localized to Xp21. These data suggested that this deletion at Xp21 resulted in a loss of a large piece of DNA that contained all the disease genes leading to his phenotype, including the gene for DMD. Dr. Kunkel and his colleagues surmised that if the DNA missing from the patient's deleted X chromosome could be isolated from the DNA of a normal individual, one would have the 10 million bp that included the sequence for the DMD gene. In this way, a DNA library containing the DNA sequences on the X chromosome that corresponded to the deleted area of X in the patient was constructed. The DNA fragments in the library were then used as probes in Southern blotting with DNA isolated from normal individuals and the patient with DMD, in a search for those that hybridized to normal DNA but not to DNA from the patient. Any such fragment (which could hybridize to normal DNA but not to the patient's DNA) would represent a DNA sequence from the 10 million bp deleted region and would thus represent a potential candidate gene sequence for the DMD gene. Of the clones isolated and tested, one clone of 180 bp was found to be missing not only in the patient but also in 10% of other male patients affected with only DMD. The DNA fragment in this clone was then used to isolate a larger DNA fragment, which resulted in the isolation of a genomic DNA fragment of 140,000 bp believed to contain sequences from the DMD gene. DNA isolated from patients with different deletions covering the DMD gene were tested with the cloned DNA fragment in an attempt to localize the position of the cloned fragment within the DMD gene. This analysis showed that the cloned fragment was located in the middle of the gene—a gene that turned out to span 2.5 million bp of DNA and to represent one of the largest human genes known. Remaining to be demonstrated, however, was whether the cloned DNA fragment represented the coding sequence for gene for DMD. To use the cloned DNA fragment to investigate the function of the DMD gene product, the DNA fragment must contain exon sequences: sequences of DNA that code for amino acids in the final gene product. Exons were present, and a mRNA of 14,000 bp that contained an open reading frame of 11,000 bp and coded for a protein of 3600 amino acids was isolated.

Identification of the Function of the Gene

With the amino acid sequence and nucleotide sequence already known for many proteins and genes and presently stored in data banks, often one can obtain information about an unknown protein by looking for sequence homology to other known proteins. When this was done with the putative DMD protein sequence, homology was found to two cytoskeletal proteins: alpha-actinin and spectrin. The putative DMD protein was isolated and used to raise antibodies in rabbits. Fluorescent-tagged antibodies specific for the putative DMD gene product were used in a search for the protein in muscle tissue from biopsy samples. These experiments revealed that the DMD protein, now known as dystrophin, was present in the muscle of normal individuals but absent in patients with Duchenne muscular dystrophy. With the isolation of the DMD gene and the protein dystrophin, molecular techniques are now available for the rapid and specific diagnosis of DMD.

For more information see Hoffee PA: *IMS: Medical Molecular Genetics*. Fence Creek, pp. 150–156.

Directions: For each of the following questions, choose the **one best** answer.

Questions 1–5

The following questions refer to the pedigree and table below. The pedigree is from a family in which several individuals are affected with an inherited disease. Affected individuals are indicated as filled symbols. There are no known sporadic cases of the disease, and the disease shows 100% penetrance.

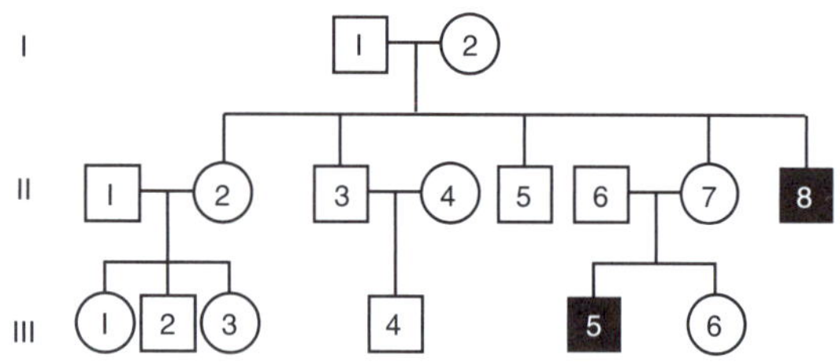

The table below presents four sets of results using SSCP analysis to detect the normal allele (*A*) and the mutant allele (*a*) of the gene.

Pedigree Member	Set 1	Set 2	Set 3	Set 4
I-1	*A*	*A*	*A/a*	*A/a*
I-2	*A/a*	*A*	*A/a*	*A/a*
II-1	*A*	*A*	*A*	*A/a*
II-2	*A/a*	*A*	*A/a*	*A/a*
II-3	*A*	*A*	*A*	*A*
II-4	*A*	*A*	*A*	*A*
II-5	*A*	*A*	*A*	*A*
II-6	*A*	*A/a*	*A*	*A/a*
II-7	*A/a*	*A*	*A*	*A/a*
II-8	*a*	*a*	*A/a*	*a*
III-1	*A*	*A*	*A*	*A*
III-2	*A*	*A*	*A*	*A*
III-3	*A*	*A*	*A*	*A/a*
III-4	*A*	*A*	*A*	*A*
III-5	*a*	*a*	*A/a*	*a*
III-6	*A*	*A/a*	*A*	*A*

1. Which data set in the table is consistent with an autosomal recessive disease?

(A) Set 1

(B) Set 2

(C) Set 3

(D) Set 4

(E) None of the data sets

2. Which data set in the table is consistent with an X-linked recessive disease?

(A) Set 1

(B) Set 2

(C) Set 3

(D) Set 4

(E) None of the data sets

3. Which data set in the table is consistent with an autosomal dominant disease?

(A) Set 1

(B) Set 2

(C) Set 3

(D) Set 4

(E) None of the data sets

4. Restriction endonucleases are best characterized by which of the following statements?

(A) They are enzymes that cleave RNA at specific sites.

(B) They are enzymes that are used in the cloning of DNA molecules.

(C) They are required for recombination between sister chromatids.

(D) They catalyze the formation of phosphodiester bonds.

(E) They catalyze the unwinding of DNA by breaking hydrogen bonds between two DNA strands.

5. A recombination value of 1%, where theta = 0.01, in genetic distance corresponds approximately to which physical distance listed below?

(A) 1000 bp

(B) 10,000 bp

(C) 100,000 bp

(D) 1,000,000 bp

(E) 10,000,000 bp

Directions: The groups of questions below consist of lettered choices followed by several numbered items. For each numbered item, select the appropriate lettered option with which it is most closely associated. Each lettered option may be used once, more than once, or not at all.

Questions 6–9

The following questions pertain to the pedigree below from a family in which two affected siblings have cystic fibrosis (CF).

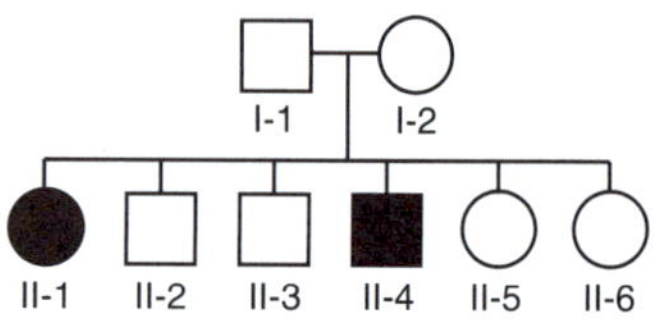

The family wishes to know which members of the family are carriers of CF and have agreed to be genotyped at three polymorphic marker loci known to be very near the CF gene. The genetic map of these loci is shown below.

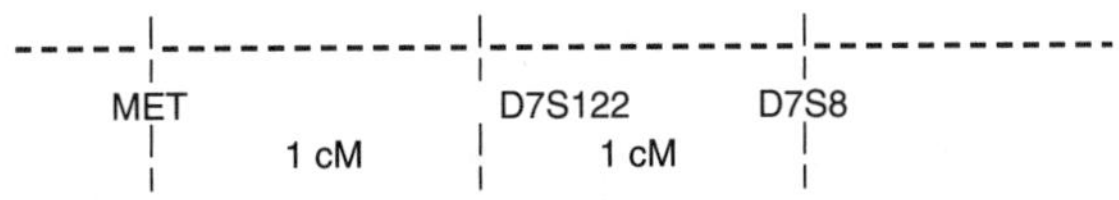

The following genotypes were identified for each individual. Note that a hyphen in the answers denotes a deletion of the marker.

	I-1	I-2	II-1	II-2	II-3	II-4	II-5	II-6
Met	a,c	b,c	c	a,c	b,c	b,c	a,b	a,c
D7S122	b,c	a,c	a,c	a,b	c	c	b,c	a,b
D7S8	a,c	a,b	a,b	b,c	a	a,b	a,c	b,c

6. The most likely haplotypes for individual II-2 are which of the following?

(A)	(B)	(C)	(D)	(E)
c a	c a	a c	a c	None of these
b a	b a	b a	b a	
b c	c b	b c	c b	

7. The most likely haplotypes for individual II-4 are which of the following?

(A)	(B)	(C)	(D)	(E)
b c	b c	b c	b c	None of these
c c	c c	c-	-c	
b a	a b	b a	b a	

8. Which of the following lists all of the individuals who are likely to be heterozygous carriers of CF in this pedigree?

(A) II-2, II-3, II-5, II-6

(B) II-2, II-3, II-6

(C) I-1, I-2, II-2, II-3, II-5, II-6

(D) I-1, I-2, II-2, II-6

(E) I-1, I-2, II-2, II-3, II-6

9. Given the data in this pedigree, which of the following statements is true?

(A) There are no recombination events; thus, the disease gene cannot be placed relative to these markers.

(B) There is a single recombination event, but the event does not allow placing the disease gene relative to these markers.

(C) There is a single recombination event, and the event places the disease gene somewhere to the left of the D7S122 marker on the genetic map given above.

(D) There is a single recombination event, and the event places the disease gene somewhere to the right of the D7S122 marker on the genetic map given above.

(E) There are multiple recombination events, and the disease does not appear to be linked to these markers.

Questions 10–14

The following questions pertain to the following pedigree of family in which the son II-2 has DMD. His two sisters want to know whether they are carriers of the disease. Below the pedigree are the electrophoreograms of the four polymorphic loci (5′*DysII*, *STR*45, *STR*49, and 3′*CA*) distributed throughout the dystrophin gene and indicated on the given map.

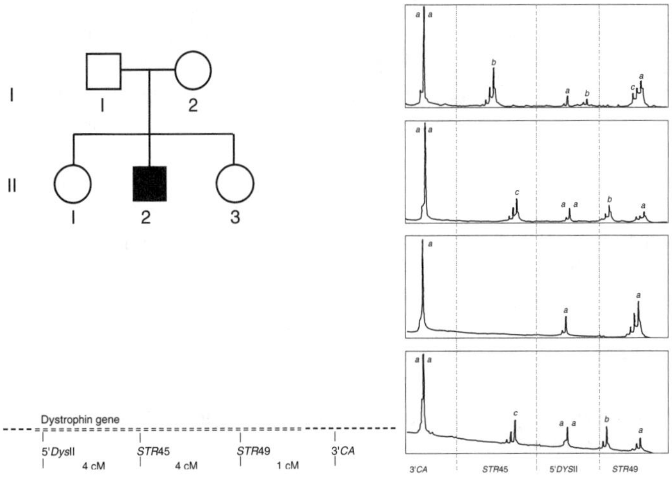

10. The correct haplotype(s) for individual II-2 is

(A)	(B)	(C)	(D)	(E)
a a	a	a	a a	a a
-	a	-	a -	a a
a a	a	a	a a	a -
a a	a	a	a a	a a

11. The correct haplotype(s) for individual I-1 is

(A)	(B)	(C)	(D)	(E)
a	a	b	a a	a a
-	c	b	- c	c c
a	b	c	b b	b b
a	a	a	a a	a a

12. The correct haplotype(s) for individual 1-2 is

(A)	(B)	(C)	(D)	(E)
a	b	a b	a b	a b
b	-	b b	- b	b -
c	c	a c	a c	a c
a	a	a a	a a	a a

13. Which of the following lists all of the individuals who are most likely female carriers?

(A) Only I-2

(B) I-2 and II-1

(C) I-2 and II-3

(D) I-2, II-1, and II-3

(E) Not enough data to determine carrier status.

14. A serum creatine kinase level was determined for II-3. The results were found to be within the normal limits. Which of the following statements is true?

(A) The results increase her carrier risk substantially.

(B) The results decrease her carrier risk substantially.

(C) The results do not significantly change her carrier risk.

(D) The results suggest she is quite athletic.

(E) The diagnosis of II-2 is probably incorrect.

PART III: ANSWERS AND EXPLANATIONS

1–3. **The answers are: 1-D, 2-A, 3-E.**

For an autosomal recessive disease, both parents are carriers of the defective allele, and an affected individual is homozygous for the defective allele. The carrier parents (*A/a*) as shown in set 4 are I-1, I-2, II-6, and II-7. The two affected individuals are II-8 and III-5, both homozygous for the defective allele. For an X-linked recessive disease, mothers are carriers of the defective allele, and sons are generally affected because they have a single copy of the X chromosome, which comes from the mother. Set 1 indicates that the mother I-2 is a carrier (*A/a*) and her son II-8 is affected and has the defective allele (*a*). The mother II-7 is also a carrier (*A/a*), and her son III-5 is affected and has the defective allele. All sons who received the *A* allele are unaffected with the disease. There is no data set consistent with an autosomal dominant disease. In that case, one would expect that all individuals carrying the *a* allele would be affected with the disease, and they are not.

4. **The answer is B.**

Restriction endonucleases cleave DNA at specific sequences to give either blunt-ended fragments or fragments with single-strand overhangs. Two fragments with complementary single-strand overhangs can be ligated together to give a chimeric DNA molecule. This ligation of two different DNA molecules is the basis of forming recombinant DNA molecules. Restriction endonucleases do not cleave RNA and are not involved in recombination. Polymerases catalyze the formation of phosphodiester bonds, and DNA helicases break hydrogen bonds.

5. **The answer is D.**

A theta of 0.01% is equal to approximately 1,000,000 bp.

6–9. **The answers are: 6-D, 7-A, 8-E, 9-D.**

To answer the questions, it is necessary to set phase for each of the individuals in the pedigree. Because CF is an autosomal recessive disease, all individuals will have two alleles at each marker locus, unless there is a deletion. If only one genotype is given, a person either is homozygous, in which case both alleles are the same, or has a deletion of one allele (noted as a hyphen). In the given genotypes, no individual has a deletion. Thus, when a single allele is given, the person is homozygous at that locus. (For example, individual II-1 has a *c* at Met, therefore is homozygous, or *c c*.) To set phase for an autosomal recessive disorder, phase is set in the parents by using an informative child. One set of alleles comes from the mother and one set from the father. Thus, we can derive the alleles of the parent by subtracting the proper allele of the informative child. Once the phase of the parents is set, the phase for each of the children can be set. If all of the children are consistent with the phase of the parents, you have set phase correctly. If there is an inconsistency, you must reset phase using another child. If phase is set properly, either no or only one recombinant is expected if the markers are close to the disease gene. However, if a recombination event is found, it can be used to determine the position of the gene mutation relative to the markers. To set phase, start with an informative child such as II-3, who is homozygous at two of the three loci, or *b/c, c/c, a/a*. One chromosome is *b c a* and the other *c c a* (see figure below). By subtracting the alleles of II-3 from each of the parents' alleles, we determine that the father must be *c c a* on one chromosome and *a b c* on the other. The mother must be *b c a* on one chromosome and *c a b* on the other. The other children are then set in phase using the phase set for the mother and the father. All of the children are consistent with the phase of the parents except the affected son II-4. His genotype is *c c a* on one chromosome (coming from the father) and *b c b* on the other. This chromosome containing *b c b* came from the mother and has a recombination event between D7S122 and D7S8 to give the genotype *b c b*. This places the disease gene somewhere to the right of the D7S122 marker on the genetic map, between D7S122 and D7S8. Both individuals in this pedigree with *b* in the third locus from the mother and *a* in the third locus from the father are affected with the disease (II-1 and II-4). Individuals with either *c c a* from the father or *c a b* from the mother are heterozygous carriers of CF. II-5 is not likely to be a carrier of CF because she did not receive the affected allele from either parent.

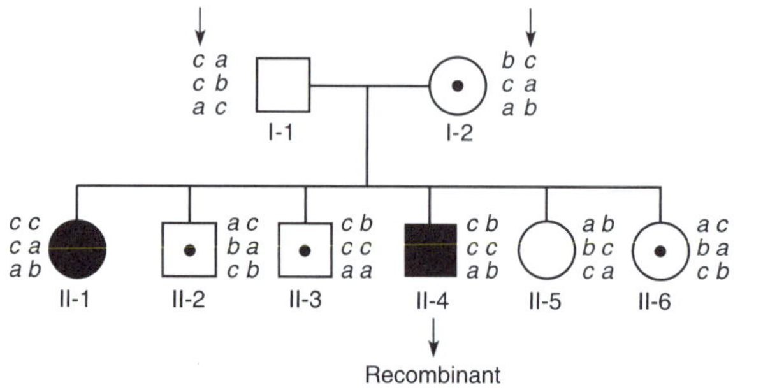

10–14. **The answers are: 10-C, 11-B, 12-D, 13-D, 14-C.**

DMD is an X-linked recessive disease affecting mostly males. To set phase in an X-linked recessive disease, one uses an affected male. Phase is already known in a male, because males have only one dystrophin gene, and therefore the single allele that a male has at each position determines his haplotype. Write the allele data from the graph next to each pedigree member in this order: 5′*DysII-STR45-STR49-3′CA*. In the given data, the male, II-2, has an *a - a a*, where the hyphen denotes a deletion of the marker. According to these data, the mother, I-2, must be *a - a a* on one chromosome and thus *b b c a*, on the other chromosome. You have now set phase for the mother. The two daughters are as follows: II-1 receives *a - a a* from the mother and thus *a c b a* from the father, making her a carrier. II-3 receives *a - a a* from the mother and thus *a c b a* from the father. She is also a carrier. The father, I-1, must be *a c b a*. All the females shown in the pedigree are carriers of the disease gene, *a - a a*. Even though II-3 has a normal serum creatine kinase level, her carrier risk based on her genotype is not changed.

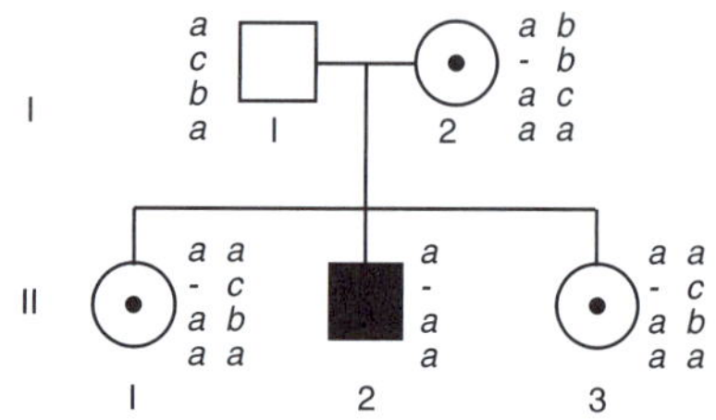

PART IV
Clinical Aspects of Human Genetics

52 Chromosomal Abnormalities I

Numerical Abnormalities

A Numerical chromosomal abnormalities

Condition	*Notation*
Normal	
Normal female	46,XX
Normal male	46,XY
Numerical abnormalities	
Triploidy	69,XXX; 69,XXY; 69,XYY
Tetraploidy	92,XXXX
Monosomy for 13	45,XX,–13
Trisomy for 21 (Down syndrome)	47,XX,+21
Mosaic for 21	47,XX,+21/46,XX

B Origin of triploid conceptions

Mechanism	*Ratio (Paternal: Maternal)*	*Frequency (%)*
Dispermy: fertilization of a normal haploid egg by two individual sperm	2:1	40
Diandry I: fertilization of a normal haploid egg by a diploid sperm because of a meiosis I error	2:1	15
Diandry II: fertilization of a normal haploid egg by a diploid sperm because of a meiosis II error	2:1	N/A
Digyny I: a single haploid sperm fertilizing a diploid egg because of a meiosis I error	1:2	7.5
Digyny II: a single haploid sperm fertilizing a diploid egg because of a meiosis II error	1:2	12.5
Noninformative	?	25

Note: N/A = not applicable.

C Nondisjunction during meiosis

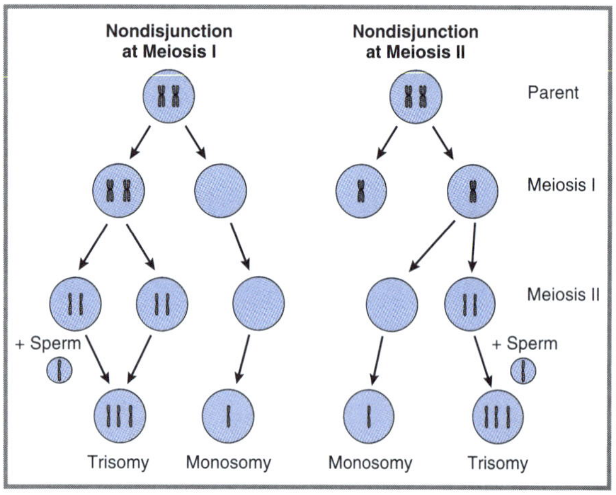

D Mechanism for uniparental disomy

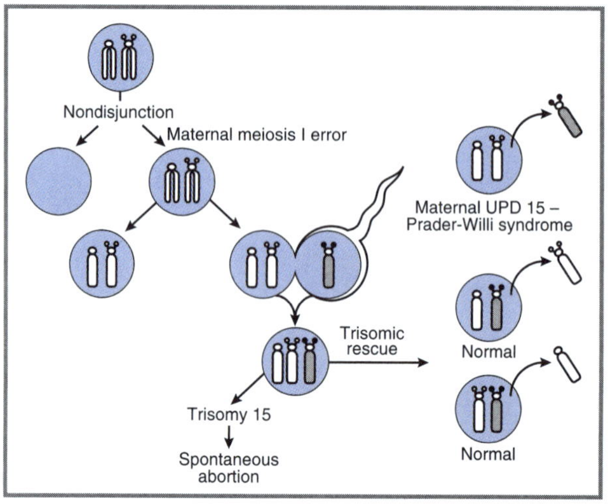

OVERVIEW

With the availability of chromosome banding techniques and FISH (see Chapter 4), it is possible to recognize changes in the structure or the number of normal chromosomes found in a metaphase cell. Chromosome abnormalities are common in humans, affecting as many as 7.5% of all conceptions. Most of these abnormal conceptions result in spontaneous abortions, reducing the live-birth frequency of chromosome abnormalities to about 0.5%, or 1 in 200 births. Chromosome abnormalities are classified as either numerical or structural. Numerical abnormalities involve a cell containing an abnormal number of normal chromosomes. Structural abnormalities involve a cell with one or more abnormal chromosomes that contain a structural rearrangement. Even cells that appear to have a normal karyotype by chromosome-banding analysis may be abnormal. For example, sometimes all of the chromosomes in a diploid cell are derived from one parent (uniparental diploidy), or sometimes both copies of a single chromosome pair are derived from one parent (uniparental disomy [UPD]). In these cases, a dramatic effect on the phenotype of the offspring can occur even though the karyotype appears normal (see Chapter 56).

Numerical Chromosome Abnormalities

Numerical chromosome abnormalities occur when the normal human chromosome number of 46 is altered, resulting in euploidy (polyploidy), aneuploidy, or mosaicism (Figures **A** and **B**). When a cell has balanced extra chromosome sets in multiples of n of 23 chromosomes, it is described as being polyploid or euploid. Cells that have a triploid (3n) number of chromosomes are identified as 69,XYY, 69,XXX, or 69,XXY. Triploidy (3n) can result from fertilization of a single egg by two sperm (dispermy) or from a defective meiosis when a diploid gamete is produced. Tetraploidy (92 chromosomes, or 4n) also occurs and is usually a result of a defect during the first zygotic division. Both triploidy and tetraploidy are lethal and generally result in spontaneous abortion of the fetus.

Aneuploidy and Nondisjunction

Aneuploidy is the term used to describe a cell in which an individual chromosome of one pair is found missing (monosomy) or a chromosome is present as an extra copy (trisomy). Monosomies, unless they involve a sex chromosome, are almost always lethal, as are most trisomies. Trisomies are common and are found in approximately 50% of all chromosomally abnormal embryos. Trisomies for all autosomes have been reported, most of which result in spontaneous abortion. However, three autosomal trisomies occurring in humans are compatible with life. They include trisomy 21 (Down syndrome), trisomy 13 (Patau syndrome), and trisomy 18 (Edwards syndrome). Trisomy 21, the first trisomy associated with human disease, was described by Lejuene in 1959, when he identified three copies of chromosome 21 in patients with Down syndrome. Monosomy and trisomy arise during either meiosis or mitosis by the process of nondisjunction, in which the two sister chromatids or two homologous chromosomes fail to separate during anaphase. Both meiosis I and meiosis II can give rise to nondisjunction, as diagramed in Figure **C**. If nondisjunction occurs during meiosis I or II, one of the gametes produced will be missing one chromosome, whereas the other will have an additional copy of the chromosome. Fertilization by a normal sperm then gives rise to either trisomy or monosomy in the offspring. The frequency of meiotic nondisjunction increases with increasing maternal age, which is reflected in the increased risk in older women of having a child with Down syndrome (trisomy 21). When nondisjunction occurs during mitosis, the offspring produced is usually a mosaic and will have two or more cell lines with a different chromosome complement. Usually a normal diploid cell line and a cell line with a chromosome abnormality are present. If mosaicism occurs early in development, multiple tissues may be affected.

Uniparental Disomy

Viable but genetically defective fetuses are produced even if only one pair of chromosomes is derived totally from either the mother or the father (UPD). UPD can occur during trisomic rescue, in which a cell with a trisomy loses one of the extra chromosomes (Figure **D**). The need for a single copy of each chromosome—one from the mother and one from the father—for normal development is related to genomic imprinting. Imprinting is a result of functional differences in the expression of genes depending on whether the gene sequences are inherited from the mother or the father. The importance of genomic imprinting in humans has become increasingly clear with the understanding of the molecular basis for two human diseases, Prader-Willi syndrome and Angelman syndrome (see Chapter 56).

Uniparental Diploidy

In some instances, the zygote receives a normal chromosome number, but the chromosomes all come from the father (uniparental diploidy), resulting in a complete hydatidiform mole (mass). The embryo is formed, but it dies early in gestation. A complete mole, with no fetus or normal placenta present, is the result of the loss of the female pronucleus in the fertilized egg followed by the duplication of the male pronucleus. Ovarian teratomas arise when a zygote contains only maternal chromosomes. This can occur when a female germ cell that is not fertilized is activated. These zygotes represent parthenogenomes because they appear to undergo partial embryogenesis without a paternal contribution. A fetus is never present; however, tissues derived from all three cell layers can be present and may give rise to some differentiated structures such as teeth, hair, and bone.

For more information see Hoffee PA: *IMS: Medical Molecular Genetics*. Fence Creek, pp. 86–89, 194–205.

53 Trisomy 21 (Down Syndrome)

A Features of Down syndrome in a neonate

Flat facial profile
Poor Moro reflex
Hypotonia
Upward-slanting palpebral fissures
Redundant loose neck skin
Hyperextensibility of joints
Dysplasia of pelvis
Simple, small round ears
Dysplasia of midphalanx of fifth finger
Single palmar crease

C Robertsonian translocation

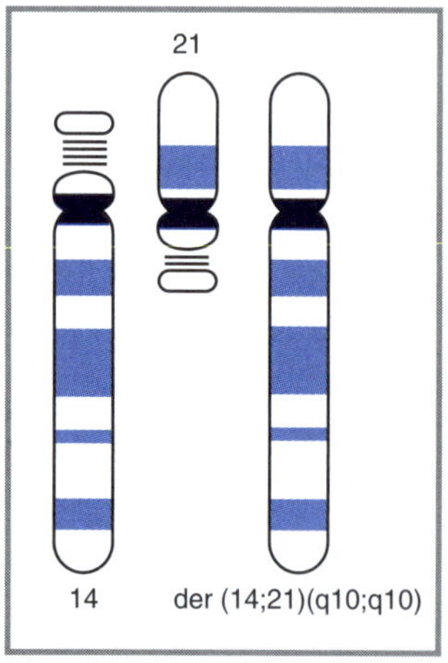

B Features of Down syndrome

D Gametes formed by a Robertsonian translocation carrier

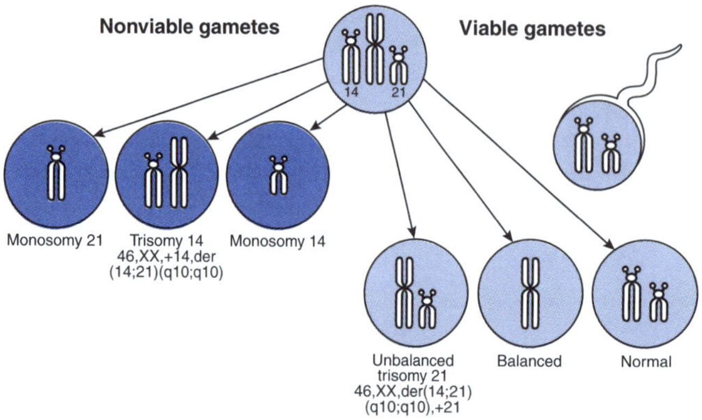

OVERVIEW

The most common chromosomal abnormality in newborns is Down syndrome, or trisomy 21. Approximately 1 in 800 children born in the United States has Down syndrome. The risk of trisomy 21 increases with the age of the mother; by the age of 35, the risk increases to 1 in 350.

Features of Down Syndrome

Down syndrome can be recognized in the neonate by the features listed in Figure **A** and shown in Figure **B**. The newborn has hypotonia and a characteristic facial appearance. Approximately 85% of infants with Down syndrome die before 1 year of age; the leading cause of death is congenital heart disease. With improved treatment and early recognition of the disease, more children now survive, and some individuals live to 50 years of age or longer. Moderate-to-severe mental retardation is characteristic of children with Down syndrome. Additionally, the children tend to be short, are obese in childhood, and have impaired hearing, eye disorders, impaired cellular immunity, and often leukemia. Alzheimer-like dementia becomes apparent by age 40, and 75% have symptoms of Alzheimer disease by the age of 60.

Origin of Trisomy 21

Ninety percent of Down syndrome cases result from nondisjunction in meiosis I during formation of the oocyte. Five percent of cases are paternal in origin and result from errors in meiosis I or meiosis II. These errors in nondisjunction lead to a gamete with three copies of chromosome 21 (see Chapter 52). Four percent of Down syndrome cases result from a Robertsonian translocation (Figure **C**), in which two acrocentric chromosomes (13, 14, 15, 21, 22) fuse near the centromere region of each chromosome, with the loss of both short arms. The most common Robertsonian translocations seen in Down syndrome are t(14;21) and t(21;22). Carriers of balanced Robertsonian translocations are phenotypically normal with 45 chromosomes, but they have an increased risk of having a child with an unbalanced chromosome. The karyotype of a balanced carrier is 45,XX, der(14;21)(q10;q10), where the derivative chromosome replaces the two missing acrocentric chromosomes, 14 and 21. The loss of the short arms from the acrocentric chromosomes does not lead to an abnormal phenotype.

Gametes Formed by a Robertsonian Translocation Carrier

During meiosis, six potential gametes can be formed by a Robertsonian translocation carrier, only three of which are viable (Figure **D**). The three viable gametes include one normal and one balanced translocation carrier and one unbalanced with both the derivative chromosome and a normal 21. After fertilization of the unbalanced gamete by a normal sperm, a gamete is produced with a karyotype of either 46,XX, der(14;21), +21 or 46,XY, der(14;21), +21 (Down syndrome). If the Robertsonian translocation involves two number 21 chromosomes in which there is a failure in centromere division, producing an isochrome for the long arm of 21 (i.e., 21q;21q), the result is either a miscarriage or a child with Down syndrome. The progeny of a balanced 21q;21q carrier either will receive no chromosome 21, an embryonic lethal condition leading to a miscarriage, or will receive the derivative chromosome leading to trisomy 21 or Down syndrome.

Diagnosis of Down Syndrome

Down syndrome can be diagnosed prenatally by ultrasound screening or biochemical analysis of proteins in the mother's serum. Because the risk of having a child with Down syndrome increases with the mother's age, screening tests take into account the risk associated with the mother's age. Three serum markers—α-fetoprotein (AFP), unconjugated estradiol (uE_3), and human chorionic gonadotropin (HCG) hormone—along with the patient's age are used to determine a risk calculation that the fetus has Down syndrome. One of the most important parameters in marker screening is the gestational age, because the serum components in the mother change with increasing age of the pregnancy. An error of even 14 days can modify the risk factor. The triple screen or maternal serum screen is generally performed between 16 and 20 weeks' gestation. If the fetus is affected with Down syndrome, there is generally a 25% reduction of the levels of AFP and uE_3 and a two-fold increase in the level of HCG. The triple screen can detect approximately 60% of Down syndrome pregnancies. Additional prenatal tests can be used to confirm the diagnosis. They include amniocentesis, chorionic villus sampling (CVS), and percutaneous umbilical blood sampling (PUBS) (see Chapter 60). The clinical diagnosis of Down syndrome is confirmed by cytogenetic analysis. This analysis now can be done by use of interphase cells and FISH in a search for three number 21 chromosomes. Chromosome analysis of the parents is important to determine whether either parent is carrying a Robertsonian translocation. When a translocation carrier is involved, the risk that the parents will have another child with Down syndrome is significantly increased.

Current Status

Chromosome 21 is one of the smallest human chromosomes and contains approximately 1.7% of the total human haploid genome. The molecular basis for the clinical phenotype seen in trisomy 21 is still not known but must be associated with the long arm of chromosome 21. The loss of the ribosomal RNA genes located on the short arms of chromosome 21 does not lead to any clinical phenotype. Analysis of cases in which only parts of chromosome 21 are present in the triploid state have implicated a region, q22.1–q22.3, as the area associated with the physical phenotype seen in affected individuals. A 2-Mbp region containing 21q22.2 has been studied in transgenic mice and appears to be associated with the learning defects that occur in Down syndrome patients. Currently, both the genetic and the physical maps of chromosome 21 are being intensely studied in the hope of identifying all the genes involved in Down syndrome.

For more information see Hoffee PA: *IMS: Medical Molecular Genetics*. Fence Creek, pp. 191–202, 300–302, 323–324.

54 Trisomy 18 (Edwards Syndrome) and Trisomy 13 (Patau Syndrome)

A Frequency of viable trisomies in 100,000 conceptions

Outcome	*Conceptions*	*Abortion Spontaneous*	*Live Births*
Trisomy 21	450 (~1/200)	345 (77%)	105 (~1/800)
Trisomy 18	200 (1/500)	190 (95%)	10 (1/8500)
Trisomy 13	195 (1/500)	190 (97.5%)	5 (1/17,000)
Total	845	725 (85%)	120 (~1/700)

B Features of trisomy 18

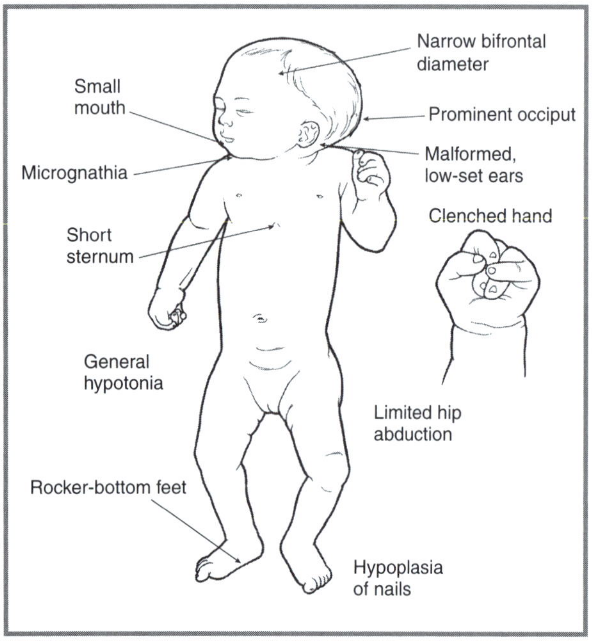

C Diagnosis of trisomy 18

Ultrasound Features of Trisomy 18
Severe, early onset IUGR
Congenital heart lesions
Diaphragmatic hernia
Polyhydramnios caused by esophageal atresia
Choroid plexus cyst
Omphalocele
Renal abnormalities
Absence of corpus callosum
Clenched fists

D Features of trisomy 13

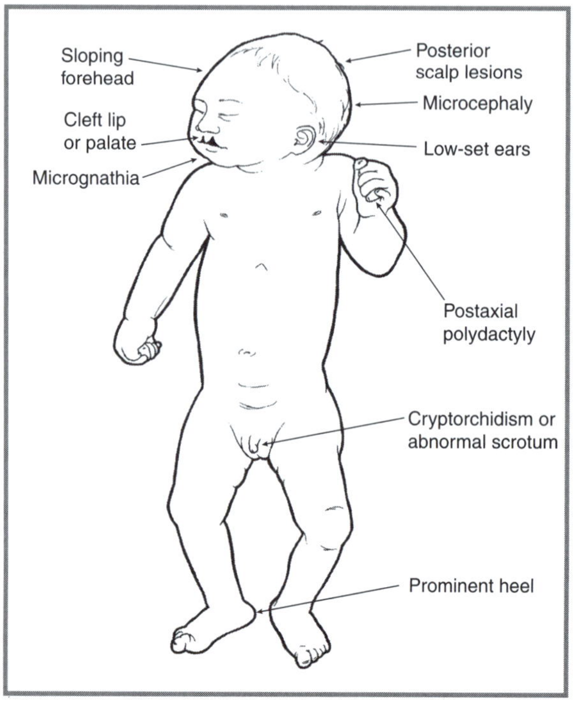

OVERVIEW

Only two other autosomal trisomies, in addition to trisomy 21, allow a fetus to survive to term (Figure **A**). They are trisomy 18 and trisomy 13. Trisomy 18, also known as Edwards syndrome, occurs in approximately 1 in 8500 births but in as many as 1 in 500 conceptions. The increased use of ultrasound now allows the detection of malformations associated with trisomy 18, accounting for the significant increase in the number of cases diagnosed prenatally in the second and third trimesters. Trisomy 13, known as Patau syndrome, is very rare, occurring in only 1 in 17,000 to 25,000 live births. Ultrasound is also used to diagnose trisomy 13 prenatally.

Features of Trisomy 18

Infants with trisomy 18 are small at delivery and display severe hypotonia and impaired neurologic development. The typical facial appearance (Figure **B**) includes a narrow head, open metopic sutures, small ears, a prominent occiput, and a receding jaw. In addition, clenched fists with the second and fifth digits overlapping the third and fourth are common, as are the appearance of rocker-bottom feet with prominent calcanei, a short sternum, and cardiac and renal anomalies. The cardiac malformations, especially the ventricular septal defects, are usually the cause of the high mortality rate in trisomy 18 infants. Because only a small number of infants (approximately 5%) survive past the neonatal period, little information is available on their development. Some children do survive into their 20s, but they have severe mental retardation and a delay in both motor and speech development. Most cases of trisomy 18 are a result of an extra chromosome 18. A few cases result from translocations. The extra chromosome 18 is maternal in origin in 90% of cases and a result of nondisjunction in meiosis II.

Diagnosis of Trisomy 18

The risk of having a child with trisomy 18 increases with maternal age and can reach an incidence of 1 in 130 by the maternal age of 43 years. Certain major malformations associated with trisomy 18 (Figure **C**) can be detected by ultrasound in the second and third trimesters. Maternal serum markers also can be used to detect approximately 80% of fetuses with trisomy 18. All three markers, AFP, uE_3, and HCG, are usually reduced below normal. Confirmation of a diagnosis of trisomy 18 by cytogenetic analysis is important so that the parents can be informed and immediate medical attention can be available at the time of birth.

Features of Trisomy 13

The phenotype of a fetus with trisomy 13 includes holoprosencephaly with incomplete development of the forebrain and olfactory and optic nerves, facial clefting, and anophthalmia, synophthalmia, or microphthalmia Figure **D**. Additional abnormalities in trisomy 13 fetuses are congenital heart defects, renal abnormalities, omphalocele, and polydactyly. Only 15%–20% of infants with trisomy 13 survive the first year of life; 44% die within the first month, and 70% die within 6 months. Death is usually the result of congenital heart defects. A few individuals survive into adulthood, but they suffer from seizures and may be deaf and blind.

Diagnosis of Trisomy 13

The characteristic phenotype associated with trisomy 13 makes it easily diagnosed by ultrasound. Once a trisomy 13 fetus is suspected, the diagnosis is confirmed by cytogenetic analysis of fetal cells obtained by amniocentesis, CVS, or PUBS (Chapter 60).

For more information see Hoffee PA: *IMS: Medical Molecular Genetics.* Fence Creek, pp. 203–205.

55 Chromosomal Abnormalities II

Structural Abnormalities

A Structural abnormalities

Structural abnormalities	
Reciprocal translocation	46,XX,t(1;12)(p35;p12)
Robertsonian translocation	45,XX,der(14;21)(q10;q10)
Terminal deletion	46,XY,del(5)(p15)
Interstitial deletion	46,XX,del(2)(q31q33)
Ring chromosome	46,XX,r(7)(p21q34)
Pericentric inversion	46,XX,inv(6)(p12q23)
Paracentric inversion	46,XY,inv(8)(q22q24)
Isochromosome	46,X,i(X)(q10)
Dicentric	46,X,dic(Y)(q13)

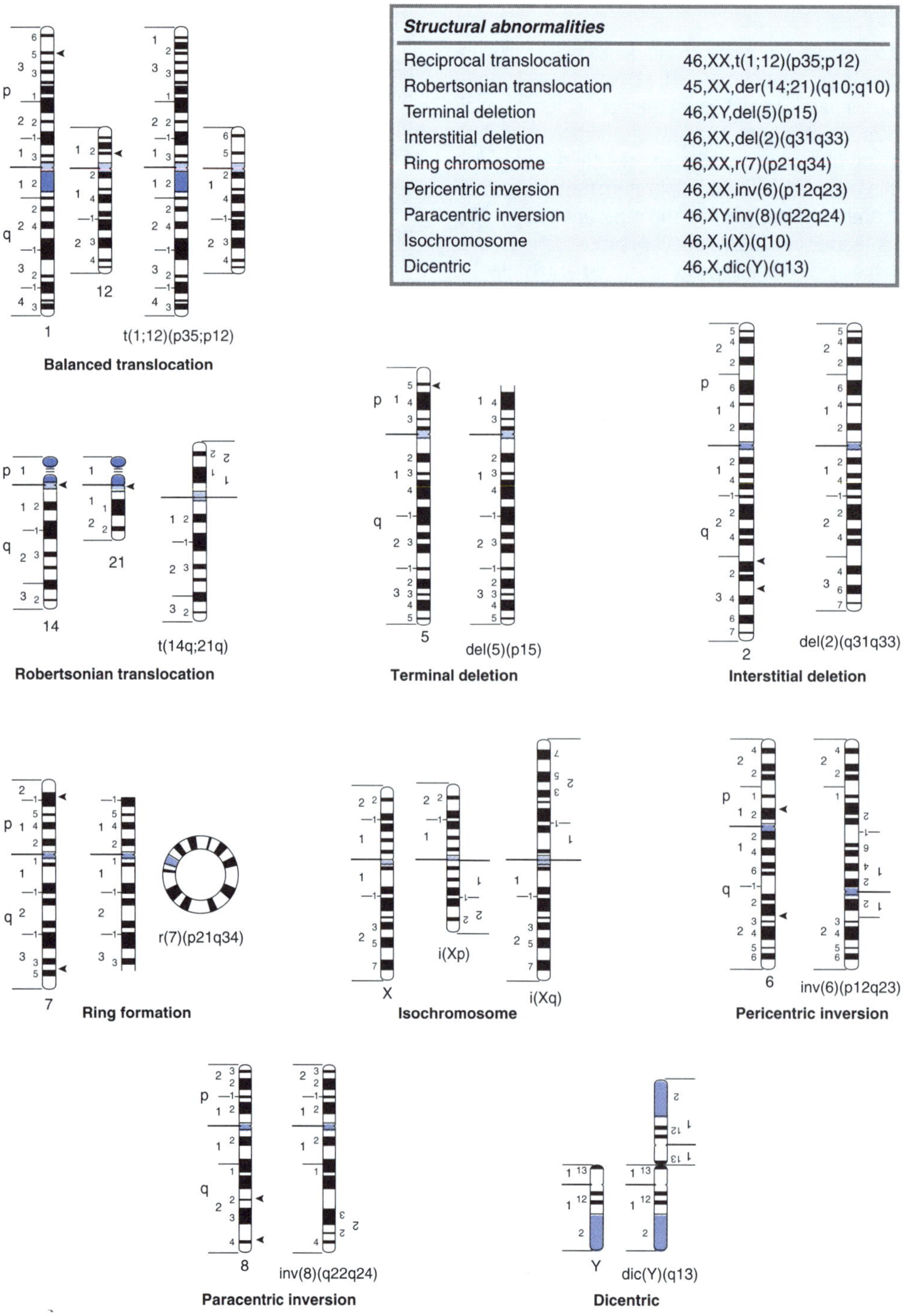

OVERVIEW

Structural chromosomal abnormalities occur at a very high frequency (1 in 500 births) in the human population. All structural rearrangements of chromosomal material come from the breakage of chromosomes and the rejoining of the fragmented ends. Normal chromosomes contain telomere sequences at the ends of each chromatid to prevent fusion of one chromosome with another or with itself. When a chromosome breaks, a free end or ends are produced. If these free ends are not repaired by rejoining to another free end, a signal to the damage control systems may eventually lead to the death of the cell. The rejoining of the ends of fragmented chromosomes is not always accurate, in which case a structural rearrangement of a chromosome results. However, if the rearrangement does not alter the content or the expression of the genetic material, no disease phenotype will be observed. Nearly all possible structural rearrangements have been seen in human karyotypes and fall into one of the following classes.

Translocations

A translocation occurs when two nonhomologous chromosomes exchange genetic material. Translocations require that at least one break occur on each of the two chromosomes. Reciprocal translocations are balanced exchanges between two chromosomes, where no genetic material is lost (Figure **A**). The chromosomes carrying the translocations are referred to as derivative chromosomes (der). Individuals with a balanced translocation will be phenotypically normal; however, their offspring may express a genetic defect if they do not receive a normal complement of genetic material. A Robertsonian translocation is the result of the fusion of two acrocentric chromosomes (i.e., chromosomes 13, 14, 15, 21, 22) at or near the centromeres. When this fusion occurs, the short (p) arms are lost from both chromosomes. This loss of the short-arm material from the acrocentric chromosomes is usually not lethal, because this part of the chromosome contains either repetitive sequences or rRNA genes that are present in many copies.

Deletions

Deletions result when a break occurs within a chromosome and genetic material is lost. Large deletions (greater than 4000 kb) can be detected by G banding as a shortened chromosome with missing bands. Terminal deletions result in the loss of genetic material at either end of a chromosome and are indicated as pter or qter. The extent of a deletion and the genes involved determine the viability and the phenotype of an individual carrying the chromosome with the deletion. Also seen are interstitial deletions, in which a chromosome breaks at two sites and the material between the breaks is deleted. Small deletions, known as microdeletions, are not always visible by banding techniques and can be detected only by DNA probes.

Ring Chromosomes

Breakage of a single chromosome at the ends of the short and the long arms with loss of the telomere sequences can result in the formation of a ring chromosome, in which the two ends of the chromosome fuse together. The presence of a centromere will allow the ring chromosome to proceed through cell division. However, cell division sometimes proceeds with difficulty, in which case a cell without the ring (monosomy) is produced.

Inversions

Inversions occur when a chromosome breaks at two positions and can involve the centromere (pericentric inversion) or not involve the centromere (paracentric inversion). Both breaks can occur on the same arm of the chromosome (paracentric inversion), or one break can occur on each arm of the chromosome (pericentric inversion). When the broken chromosome is rejoined, the fragment is attached in an inverted order. If genetic material is not lost or gained or altered in expression by the inversion, there are no obvious effects on the phenotype of the individual. However, since the DNA sequence has been inverted on one of the homologues, interference with chromosome pairing at prophase I in meiosis I may be affected. Inefficient or defective pairing can result in deletions or duplications of the genetic material, which are later expressed in the offspring.

Isochromosomes

When division occurs along the wrong axis of a chromosome, the resulting chromosome has two copies of one arm and is missing a copy of the other arm (i.e., an isochromosome). In most cases, an isochromosome is lethal, with the exception of an isochromosome involving the X chromosome or chromosome 21. Individuals with an isochromosome of the X chromosome resemble patients with Turner syndrome and are monosomy for one arm of X and trisomy for the other arm. Isochromosomes of the long arm of chromosome 21 are seen in some individuals with trisomy 21.

Other Structural Abnormalities

Other types of structural abnormalities seen in human karyotypes include chromosomes with duplications, chromosomes with amplifications of large segments, and chromosomes with insertions of genetic material. In addition, chromosomes with two centromeres or dicentric chromosomes are observed. Two structural chromosomal abnormalities seen in many tumor cells result from the amplification of large pieces of DNA (see Chapter 59). One chromosome abnormality of this type is the double minute (DM) chromosome. DMs are extrachromosomal DNA elements that do not contain a centromere. The other chromosome abnormality associated with the amplification of genetic material involves the presence within chromosomes of 100–1000 copies of a particular gene sequence. These amplified sequences can be detected in a light microscope after G banding and appear as large areas within a chromosome or at the ends of chromosomes that stain very lightly. This nonbanded area is referred to as a homogeneously staining region (HSR).

For more information see Hoffee PA: *IMS: Medical Molecular Genetics*. Fence Creek, pp. 89–93, 205–211.

56 Microdeletion Syndromes

Prader-Willi and Angelman Syndromes

A Microdeletion syndromes

Chromosome Location	Syndrome	Features
del(4)(p16.3)	Wolf-Hirschhorn syndrome	Dysmorphic features Mental retardation Growth retardation/microcephaly Ocular hypertelorism Cardiac defect
del(5)(p15.5)	Cri du chat syndrome	High-pitched, monochromatic cry Moderate-to-severe mental retardation Hypotonia Growth delay
del(7)(q11.2)	William syndrome	Dysmorphic features Failure to thrive Developmental delay and overly friendly Personality Supravalvular aortic stenosis Hypercalcemia/hypercalciuria Musculoskeletal abnormalities
del(8)(q24.1)	Langer-Giedion syndrome/ tricho-rhinophalangeal	Multiple exostoses Dysmorphic features Variable intelligence
del(11)(p13)	WAGR syndrome	Wilms' tumor Aniridia Genitourinary dysplasia Mental retardation
dup(11)(p11.5) paternal	Beckwith-Wiedemann syndrome	Exomphalos Macroglossia Macrosomia Hypoglycemia Hemihypertrophy/overgrowth
del(15q) (11q–13) paternal	Prader-Willi syndrome	Mental retardation Hypotonia Obesity Dysmorphic features Hypopigmentation
del(15) (q11q–13) maternal	Angelman syndrome	Mental retardation Seizures Ataxic gait and hand flapping Inappropriate laughter Hypopigmentation
del(16)(p13.3)	Rubinstein-Taybi syndrome	Mental retardation Dysmorphic features Broad thumbs and toes
del(17)(p13.3)	Miller-Dieker syndrome	Lissencephaly (agyria, pachygyria) Profound mental retardation Dysmorphic features
del(17)(p11.2)	Smith-Magenis syndrome	Mental retardation Dysmorphic features Hyperactivity Self-destructive behavior
dup(11)(p11.2)	Charcot-Marie-Tooth disease type 1A	Hypotonia Decreased reflexes Club foot
del(20) (p11.2p12)	Alagille syndrome	Chronic cholestasis Dysmorphic features Vertebral anomalies Ocular embryotoxin
del(22)(q11.2)	DiGeorge syndrome/ Velocardiofacial syndrome	Absent or hypoplastic thymus and parathyroid Cardiac malformations Dysmorphic features Mental retardation (variable) Cleft palate
dup(22q)	Cat-eye syndrome	Coloboma of the iris Anal atresia Ear abnormalities Cardiac defects

B Genetic map of 15q11–q13

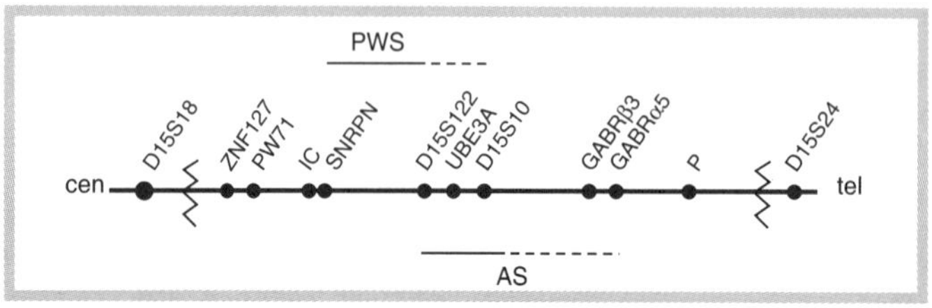

C Characteristics of Prader-Willi and Angelman syndromes

	Prader-Willi Syndrome	Angelman Syndrome
Chromosome abnormality	70% have a deletion at 15q11–13 Chromosomes 15 are both of maternal origin (25%) Imprinting error (~5%)	70% have a deletion at 15q11–13 Chromosomes 15 are both of paternal origin (2%) Imprinting error (3%)
Suggested gene products affected	Small nuclear ribonucleoprotein-associated peptide N functions in spliceosomes Other unidentified genes	Ubiquitin-protein ligase
Clinical features	Neonatal hypotonia Hypogonadism Short stature Obesity with uncontrolled appetite Mild mental retardation Hypopigmentation (fair with blue eyes) Behavioral problems	Severe mental retardation Seizures Inappropriate laughter and ataxic gait (happy puppet syndrome)

OVERVIEW

Small chromosome deletions (microdeletions) that are associated with inherited human disease syndromes are often not visible with normal cytogenetic techniques and can be detected only with the use of molecular probes. In most of these cases, multiple unrelated genes located in proximity to each other are deleted and lead to the various phenotypes associated with these microdeletion syndromes. Most syndromes are characterized by mental retardation, along with a spectrum of phenotypic abnormalities, although the severity of the abnormality depends on the size and location of the deletion. Some of the better characterized microdeletion syndromes are listed in Figure **A** and discussed below.

Cri du Chat, del(5)(p15.5); Miller-Dieker, del(17)(p13.3); Smith-Magenis, del(17)(p11.2); and DiGeorge, del(22)(q11.2) Syndromes

Cri du Chat Syndrome.
Cri du chat syndrome is characterized by a shrill cat-like cry in infancy, which disappears a few weeks after birth. The infants have low birth weight, delayed growth, and faces characterized by microencephaly, roundness, downward slanting of the eyes, strabismus, a broad nasal base, and low-set ears. These children have severe mental retardation, hypotonia, and congenital heart defects.

Miller-Dieker Syndrome.
This syndrome is characterized by an incomplete development of the brain, often with a smooth surface (lissencephaly). Facial features include a prominent forehead, a short nose with upturned nares, a protuberant upper lip, and a small jaw. Infants with Miller-Dieker syndrome have severe mental retardation, and death is common before 2 years of age. The lissencephaly seen in this syndrome appears to be caused by a deletion of a single gene (*L1S1*) within the deleted region. The function of the gene product is not known, but it may play a role in signal transduction.

Smith-Magenis Syndrome.
Smith-Magenis syndrome has an incidence of 1 in 25,000 live births and results from a deletion on chromosome 17 at p11.2. Two genes have been mapped to this area: a gene coding for a small nuclear RNA known as U3 that is involved in RNA splicing, and a gene coding for a microfibril-associated glycoprotein. The relationship of the loss of these two gene functions to Smith-Magenis syndrome has not been established. Smith-Magenis syndrome is characterized by multiple congenital anomalies, dysmorphic features, developmental delay, and mental retardation. Affected individuals show destructive behavior patterns that include head-banging, wrist-biting, and nail-picking.

DiGeorge Syndrome.
DiGeorge syndrome is characterized by hypertelorism, short philtrum, downward-slanting palpebral fissures, hypoplasia of the thymus and the parathyroid, which leads to severe hypocalcemia and seizures in affected infants. These individuals have a cleft palate, cardiac defects, and learning disabilities. Infants die of infections because of their lack of cellular immunity, or of cardiac failure. A related disorder, velocardiofacial syndrome (VCFS), is characterized by a similar phenotype and is associated with the same deletion. The incidence of these disorders is 1 in 20,000 live births. Generally, the cases tend to be mild and often are not diagnosed.

Prader-Willi and Angelman Syndromes

Approximately 70% of patients with Prader-Willi syndrome (PWS) or Angelman syndrome (AS) have been shown by DNA technology to have a common small deletion at 15(q11–q13), localizing the genes involved in these two syndromes to a specific site on chromosome 15 (Figure **B**). Molecular studies have identified the candidate gene for PWS as a gene coding for a small nuclear ribonucleoprotein-associated peptide (*SNRPN*) mainly expressed in the brain. This peptide is involved in the functioning of spliceosomes and is apparently expressed only from the parental chromosome 15. The gene involved in AS maps in proximity to the *SNRPN* gene but is a separate gene. The gene (*UBE3A*) involved in AS is expressed only from the maternal chromosome and codes for an enzyme known as ubiquitin-protein ligase. *SNRPN* and *UBE3A* are imprinted genes, with *SNRPN* expressed from the paternal chromosome 15 and *UBE3A* expressed from the maternal chromosome 15. When the paternal chromosome 15 is missing (UPD), a child develops PWS. When the maternal chromosome 15 is missing, the child develops AS. UPD accounts for approximately 25% of PWS cases and 2% of AS cases.

Characteristics of Prader-Willi Syndrome

Figure **C** summarizes the characteristics of PWS and AS. The incidence of PWS is 1 in 15,000 live births. Because the deletion is too small to detect by routine cytogenetic analysis, PWS is not usually detected by prenatal diagnosis. No phenotypic abnormalities are detected by ultrasound, although there is often a decrease in fetal movement. Infants are hypotonic at birth and have a poor suck reflex, making feeding difficult. The child appears normal until about 2 years of age, when he or she begins to have behavior problems and develops an unusual appetite that cannot be satisfied. This uncontrollable desire to eat leads to food stealing, raiding garbage cans, and eating nonfood items. A rapid gain in weight and extreme obesity result if food intake is not controlled. Families with a child affected with PWS often lock their refrigerators and kitchens and take unusual measures to prevent the child from leaving the house to obtain food. Affected individuals display uncontrollable behavior problems such as violent outbursts and temper tantrums, and they have been known to burn down their houses. PWS patients must be on a strict diet and exercise program to control their eating and weight problems. They usually have mild mental retardation and generally must live in a supervised group home.

Characteristics of Angelman Syndrome

The incidence of AS is 1 in 25,000 live births. The diagnosis of AS is made after 1 year of age with the onset of seizures. Developmental symptoms can occur between 6 and 12 months of age and include a lack of speech, motor delays, and balance disorders. Individuals with AS show characteristic "puppet-like" gait and ataxia movements. These children are severely mentally retarded, laugh frequently, and have a happy demeanor. The syndrome is often referred to as the happy puppet syndrome.

For more information see Hoffee PA: *IMS: Medical Molecular Genetics*. Fence Creek, pp. 87–89, 205–211.

57 Sex Chromosomes I

The X and Y Chromosomes and X Inactivation

A Ideogram of the X chromosome

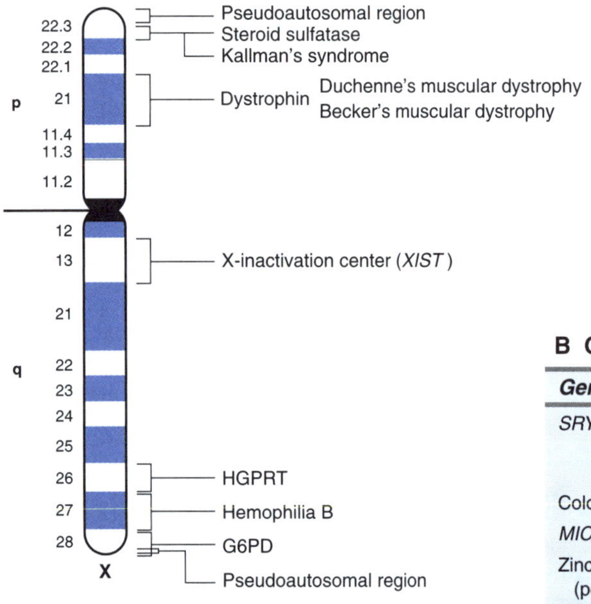

B Genes located on the Y chromosome

Genes	*Location*
SRY gene	Yp (pseudoautosomal region)
Colony-stimulating factor 2 receptor α-chain	Yp
MIC2 gene	Yp
Zinc-finger Y gene (possible role in sex determination)	Yp
Ribosomal protein S4 gene	Yp
Histocompatibility-Y antigen gene (*H-Y* gene)	Yq11*
Major gene for stature	Yq11*
Azoospermia-related gene	Yq11*
Gonadoblastoma gene	Yq11*

*Tentative assignment

C Model for the initiation of X inactivation

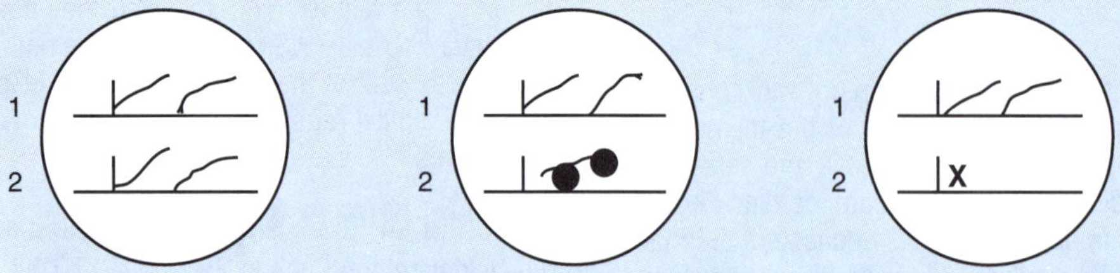

Transcription from both X chromosomes of XIST RNA

Destabilization of XIST RNA on active X chromosome by limiting factors and RNA turnover

Silencing by methylation of *XIST*gene on active X chromosome; *Cis*-acting stable XIST RNA binds to inactive X chromosome and blocks transcription

OVERVIEW

A normal human female has two copies of the X chromosome, and a normal human male has one copy of the X chromosome and one copy of the Y chromosome. A father transmits an X chromosome to all of his daughters and a Y chromosome to all of his sons, whereas a mother transmits one copy of an X chromosome to both her daughters and her sons. Because sons receive their X chromosome only from their mother, a father with an X-linked disorder cannot transmit the defective gene to any of his sons and will transmit the defective gene to all of his daughters. Even though a female has two copies of the X chromosome, she does not have twice the level of gene products coded for by the genes on the X chromosome. This dosage compensation, in which the level of X-linked gene products remains the same in both human males and females, is a result of the process known as X inactivation.

X Chromosome

The X chromosome makes up approximately 10% of the total human genome. X-linked disorders can be followed through family pedigrees; this characteristic has resulted in the mapping to the X chromosome of many genes that are associated with inherited human disorders. Only a few of the nearly 300 genes mapped to the X chromosome are indicated on the ideogram in Figure **A**. Notable among them is the gene coding for dystrophin, which is defective in children with Duchenne or Becker muscular dystrophy (see Chapter 51) and the *XIST* gene, which marks the X-inactivation center (Xic) (see below).

Y Chromosome

The Y chromosome is one of the smaller human chromosomes and can vary in size from one male to another as a result of different amounts of heterochromatin. No known phenotypic properties are associated with this variation in size. Approximately 25 genes have been mapped to the Y chromosome, some of which are listed in Figure **B**. The *SRY* gene (the testis-determining factor or sex-determining region Y) is located on the short arm of the Y chromosome adjacent to the pseudoautosomal region. The presence or absence of the *SRY* gene determines the sex of an individual. The *SRY* gene codes for a DNA-binding protein containing a zinc finger domain that may play a role in the regulation of gene expression. In the presence of the *SRY* gene product, the primitive gonads develop into testicular tissue, which then secrete the androgens necessary for the development of external male genitalia. When the *SRY* gene product is absent (i.e., in the XX female state) ovaries develop as well as female external genitalia. Thus, femaleness is due to the absence of the gene product of the *SRY* gene. If the *SRY* gene is translocated from the Y chromosome to the X chromosome, an individual with the karyotype 46,XX (who is a phenotypically normal male with small testes and a lack of spermatogenesis) or an individual with the karyotype 46,XY (who is a normal female) can result. Located on the terminal end of the short arm of the Y chromosome as well as on the short arm of the X chromosome is a region called the pseudoautosomal region. During male meiosis, the X chromosome and Y chromosome pair at these homologous regions and undergo recombination.

X Inactivation

Sex chromatin masses, known as Barr bodies, are found in the interphase nucleus of females but not in the nucleus of males. The number of Barr bodies present in an interphase nucleus is directly related to the number of X chromosomes present and is equal to one less than the total number of X chromosomes. Each Barr body represents an inactive X chromosome. Dosage compensation is maintained in female cells by having only one X chromosome transcriptionally active (i.e., the active X) and all other X chromosomes transcriptionally inactive (i.e., the inactive X). Inactivation of one of the X chromosomes occurs early in embryonic life, approximately 3 days after fertilization and is complete by the end of the first week of development. The inactivation process is thought to be random, and either the maternal or the paternal X chromosome can be inactivated. Only in extraembryonic tissue membranes is there a selection for inactivation of the paternal X chromosome. Once one of the X chromosomes has been inactivated, all daughter cells will have the same chromosome inactivated, which results in females being mosaic for X-linked genes (i.e., some cells will have one X chromosome inactivated, and some cells will have the other X chromosome inactivated). Most genes on the inactive X chromosome are transcriptionally inactive, except for the pseudoautosomal region, the steroid sulfatase gene, and the *XIST* gene (see Figure **A**). The *XIST* gene is required for X inactivation. Once inactivation is complete, the *XIST* gene is expressed only from the inactive X chromosome. The product of the *XIST* gene is an RNA molecule that functions directly as an RNA molecule and does not code for a protein, although it does undergo splicing and polyadenylation. XIST RNA transcripts bind to the inactive X chromosome and act only in *cis* (i.e., they associate only with the chromosome from which they originate). The stability of the XIST RNA appears to be the important factor in the choice of which X chromosome is to be inactivated. In Figure **C**, a putative model is presented for the initiation of X inactivation. Initially, XIST RNA is transcribed from both X chromosomes (1 and 2). As X inactivation begins, an active destabilization of the XIST RNA transcribed from the X chromosome (2) that is to remain active occurs. This destabilization is potentially a result of the presence of yet unidentified factors that are limited in quantity. Thus, only XIST RNA from one X chromosome is destabilized, while the XIST RNA transcribed from the other X chromosome (1) is stabilized and accumulates. When X inactivation is completed, the XIST gene on the active X chromosome (2) is silenced by methylation, and the XIST RNA transcribed from the inactive X chromosome (1) binds to the DNA and blocks transcription from the inactive X chromosome. In the absence of XIST RNA, genes on the active X chromosome are actively transcribed.

For more information see Hoffee PA: *IMS: Medical Molecular Genetics*. Fence Creek, pp. 211–219.

58 Sex Chromosomes II

Numerical and Structural Abnormalities

A Features of Turner syndrome

Short stature
Webbing of the neck
Low posterior hairline
Lack of secondary sexual characteristics
Primary or secondary amenorrhea
Gonadal dysgenesis

B Sex chromosome abnormalities

	Karyotype	Disorder	Phenotype
Female	45,X	Turner syndrome	Short female Delayed puberty
	46,X,i(X)(q10)	Turner syndrome	(same as above)
	46,X,X,p-	Turner syndrome	(same as above)
	47,XXX	Trisomy X	Tall female
	46,XY	Sex reversal	
		X-linked defect in testis pathway	Gonadoblastoma
		Y-linked mutation in *SRY* gene	
		X-linked androgen insensitivity	Female contours
Male	47,XXY	Klinefelter syndrome	Tall male
	47,XYY	None	Tall male
	46,XX	Sex reversal	
		Translocation of *SRY* gene to X chromosome	Normal height

C Multiple malformation syndromes with sex reversal

Disorder	Phenotypic Abnormalities	Sexual Development
Congenital adrenal hypoplasia	Normal appearance at birth Salt-wasting Accelerated skeletal maturation Rapid growth in childhood	Ambiguous genitalia in females Normal in males
Smith-Lemli-Opitz syndrome	Microcephaly Mental retardation Hypotonia Polydactyly Small nose with anteverted nares Ptosis	Ambiguous genitalia in males Hypospadias Complete failure to develop External male genitalia with mixed internal structures
Campomelic dysplasia	Dwarfism Extreme bowing of leg bones Small scapulae Craniofacial changes Defective tracheal bronchial Cartilage→respiratory distress→neonatal death	46,XY females with ovarian dysgenesis but normal uterus and fallopian tubes
Denys-Drash syndrome	Wilms tumor Aniridia Severe urogenital aberrations	Ambiguous genitalia in males Rudimentary uterus Fimbriated fallopian tubes Streak gonads
Meckel-Gruber syndrome	Encephalocele Dysplasia of kidneys Polydactyly Neonatal death	Cryptorchidism Incomplete development of external or internal genitalia

OVERVIEW

Unlike numerical chromosome abnormalities involving autosomal chromosomes, sex chromosome numerical abnormalities often have very mild or no phenotypic effect, and most cases remain undiagnosed unless a clinical problem develops. Two common human disorders caused by an abnormal number of sex chromosomes are Turner syndrome and Klinefelter syndrome. Structural abnormalities of both the X chromosome and the Y chromosome also occur and involve deletions, translocations, or structural rearrangements. Additionally, individuals may present with ambiguous genitalia and sex reversal phenotypes; examples are males with 46,XX karyotype and females with 46,XY karyotypes.

Turner Syndrome

Turner syndrome is associated with monosomy of the X chromosome and accounts for nearly 20% of all spontaneous abortions. Approximately 95% of conceptions with Turner syndrome are aborted, and it is seen in only 1 in 2500 liveborn females. Infants who survive have relatively mild phenotypes and grow into adulthood. The most common form of Turner syndrome is 45,X (approximately 50% of all cases), with the absence of the X chromosome a result of either nondisjunction in females or anaphase lag of the Y chromosome during meiosis in males. Sixty-five percent of Turner syndrome individuals retain the maternal X chromosome, although there is no maternal age effect and no increased risk for future pregnancies. Short stature and delayed puberty (Figure **A**) often indicate a potential case of Turner syndrome, and if the condition is recognized, treatment to promote growth and the development of secondary sexual characteristics can be carried out. Almost all women with Turner syndrome are infertile as a result of degeneration of their oocytes. Considering the fact that normal females have only one active X chromosome, why is any abnormal phenotype associated with Turner syndrome? One hypothesis suggests the presence of a gene on the inactive X chromosome that escapes inactivation and that has a homologue on the Y chromosome. Another possibility is that a necessary gene is expressed from both X chromosomes before X inactivation occurs in the fetus.

Sex Chromosome Abnormalities

Approximately 50% of all cases of Turner syndrome are 45,X, and the other 50% of cases are attributed to either chromosomal mosaicism or structural abnormalities of the X chromosome. The most common structural abnormality of the X chromosome is the formation of an isochromosome of the long arm, 46,X,i(X)(q10). Deletions of the short arm of the X chromosome, 46,X,X,p-, also give rise to Turner syndrome (Figure **B**). Individuals with Klinefelter syndrome have a 47,XXY karyotype and suffer from hypogonadism. The incidence of the syndrome is approximately 1 in 1000 male births. The classic features of Klinefelter syndrome manifest themselves in the adult male and include tall stature, small testes, and infertility. Infants have no abnormal phenotype, and affected individuals appear normal until puberty, when they begin to show a lack of secondary sexual characteristics. The average IQ of an individual with Klinefelter syndrome ranges from 85 to 90, and approximately 70% of 47,XXY males have developmental and learning disabilities. Approximately 50% of 47,XXY conceptions are the result of maternal meiosis nondisjunction, and 50% are the result of paternal meiotic nondisjunction.

In 47,XYY and 47,XXX syndromes, approximately 1 in 1000 liveborn males have a 47,XYY karyotype. This karyotype has no phenotypic abnormalities associated with it, and males with a 47,XYY chromosome complement are taller than those with a normal karyotype. Trisomy X (47,XXX) occurs in approximately 1 in 1000 liveborn females but usually is not associated with any phenotypic abnormalities. Some similarities to Klinefelter syndrome may be present, such as tall stature and mild learning disabilities. In general, females with trisomy X are sexually normal and produce normal offspring. Individuals who are considered to be pseudohermaphrodites are males with a 46,XX chromosome complement. They occur in approximately 1 in 20,000 live male births and are usually the result of the translocation of the *SRY* gene from the Y chromosome to the X chromosome. Individuals with the translocation have a phenotype similar to those with Klinefelter syndrome, except they have normal intelligence and are of normal height. In females with a 46,XY chromosome complement, the condition is thought to arise by several methods, including an X-linked recessive form in which a mutation in a gene required for the testis-determining pathway has occurred, and a Y-linked recessive form caused by a mutation in the *SRY* gene (see Figure **B**). Females with the 46,XY karyotype do not undergo pubertal changes because they do not secrete the necessary female hormones. They tend to develop gonadoblastoma, and gonadectomy is recommended to prevent tumor formation. A second X-linked recessive disorder, which has an incidence of 1 in 20,000, gives rise to 46,XY females who have testicular tissue that secretes androgens. Approximately 75% of these cases are a result of a mutation in the androgen receptor gene, which is located on X chromosome. This disorder, previously known as testicular feminization, is now called complete androgen insensitivity syndrome. Individuals with this syndrome appear phenotypically as normal females with breast development and female contours. They have sparse auxiliary and pubic hair, a short vagina, and some remnants of a uterus and fallopian tubes, and they are infertile.

Multiple Malformation Syndromes with Sex Reversal

Additional syndromes with sex reversal phenotypes also occur and are listed in Figure **C**.

For more information see Hoffee PA: *IMS: Medical Molecular Genetics*. Fence Creek, pp. 212–218.

59 Chromosomal Abnormalities Associated with Cancer

A Chromosome translocation alters expression of the *MYC* gene

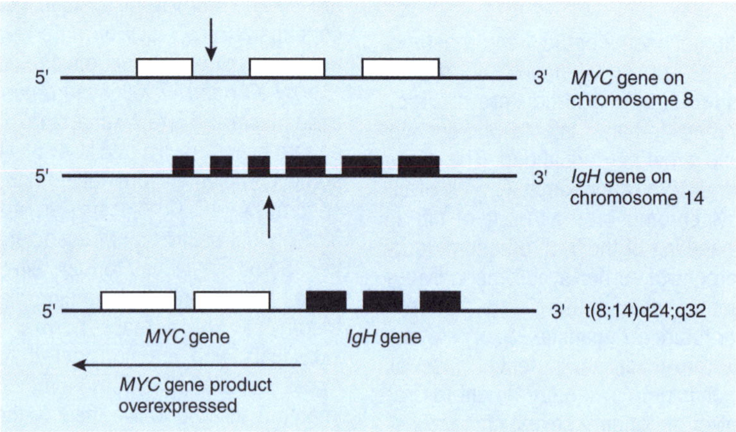

B Chromosome translocation alters the structure of the *ABL* gene

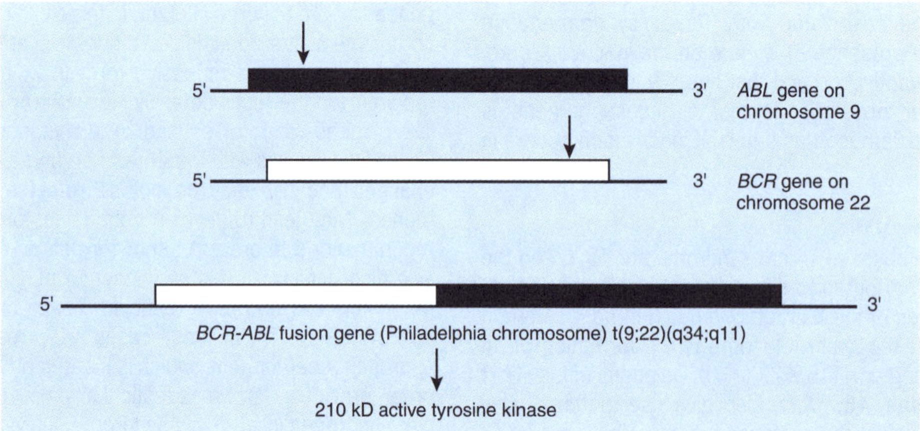

C Chromosome abnormalities in solid tumors

Tumor	*Abnormality*	*Genes involved*	
Malignant			
Synovial sarcoma	t(X,18)(p11.2;p11.2)	*SSX1,SSX2*	*SYT*
Myxoid liposarcoma	t(12;16)(q13;p11)	*CHOP*	*FUS*
Ewing's sarcoma	t(11;22)(q24;q12)	*FLI1*	*EWS*
Retinoblastoma	del(13)(q14q14)	*RB1*	
Wilms' tumor	del(11)(p13p13)	*WT*	
Benign			
Salivary gland tumor	t(1;12)(p22;q15)	*HMGIC*	
Uterine leiomyoma	t(12;14)(q15;q24)	*HMGIC*	

Note: t = translocation; del = deletion; p = short arm of chromosome; q = long arm of chromosome

OVERVIEW

Karyotypes of cancer cells show multiple abnormalities with extra chromosomes, missing chromosomes, and rearranged chromosomes. Although many of these changes are a reflection of the genetic instability of the DNA in tumor cells, some of the abnormalities are always associated with a particular type of cancer. Many types of cancers are found in humans; they include carcinomas (cancer of epithelial tissue), sarcomas (cancer of connective tissue), and leukemia or lymphomas (cancer of hematopoietic tissue). All types of malignant cells have as their hallmark uncontrollable growth and cell proliferation, which result from the activation of oncogenes or the alteration of tumor suppressor genes. Breakage of chromosomes followed by translocation can give rise to changes in gene expression and/or changes in gene structure. These quantitative and qualitative changes are now known to activate oncogenes, which have the ability to alter the growth patterns of a cell (see Chapters 21–23).

Chromosome Translocations Alter Gene Expression

An example of a translocation that alters the quantitative expression of a gene is that associated with Burkitt's lymphoma, which is common in regions of Africa where there is a high incidence of malaria. The chromosomal translocation, t(8;14) (q24;q32), is found in nearly 85% of patients with this cancer and has resulted in the translocation of the *MYC* gene from its normal location on chromosome 8 into the middle of the immunoglobulin heavy chain gene on chromosome 14 (Figure **A**). The effect of this translocation is to place the *MYC* gene under the transcription controls of an actively transcribed gene, resulting in an overexpression of the normal *MYC* gene product in B cells. The *MYC* gene product, normally a transcription factor, gives rise to inappropriate regulation of gene expression when overexpressed, leading to altered cell growth and tumor development.

Chromosomal Translocations Alter Gene Structure

An example of a translocation that alters the structure of a gene and a gene product is the translocation associated with chronic myelogenous leukemia (CML). This reciprocal translocation, t(9;22)(q34;q11), produces a small acrocentric chromosome, also known as the Philadelphia chromosome, in which part of the *ABL* gene is translocated to the breakpoint cluster region (*BCR*) on chromosome 22. The result of this translocation is the creation of a chimeric gene, part *ABL* and part *BCR* (Figure **B**). The normal *ABL* gene codes for a protein of 140 kd, whereas the fused gene now codes for a protein of 210 kd that has high tyrosine kinase activity and results in increased cell growth and tumor formation.

Chromosomal Abnormalities Associated with Solid Tumors

Increased sensitivity in cytogenetic techniques has resulted in the identification of several specific chromosomal alterations in solid tumors, including translocations, deletions, and rearrangements (Figure **C**). Retinoblastoma (RB), a childhood tumor of the retina, is caused by a mutation in the *RB* gene (see Chapter 23). A deletion at 13(q14.1–q14.2) associated with approximately 3% of RB cases aided in the identification of the *RB* gene and its eventual cloning. Additional point mutations in the *RB* gene also give rise to RB. The *RB* gene product normally acts as a tumor suppressor TF and is important for proper regulation of the cell cycle (see Chapter 23).

Gene Amplification

Overexpression of normal gene products can occur by translocations that alter the rate of expression of a gene, as discussed above for Burkitt's lymphoma. Overexpression of specific gene products in tumor cells is also associated with two aberrant chromosome types: DMs and HSRs. DMs are observed in stained karyotype as small extrachromosomal circular DNA molecules present in doublets. These chromatin structures lack centromeres and are randomly distributed to daughter cells at mitosis. DMs usually have 5–6 copies of a particular gene, and a single cell may have hundreds of copies of the DMs. The product of the gene carried on the DMs may therefore be expressed at 500 times the normal level. HSRs also contain hundreds of copies of a particular gene sequence, but they differ from DMs in that they are found as part of a normal chromosome, often attached to the ends, and are stably inherited. HSRs stain very lightly with G banding and usually are devoid of bands. Both chromosome abnormalities have been associated with late-stage neuroblastomas, in which the *MYC* gene is amplified and overexpressed, as well as in some types of breast cancer in which the *ERBB2* and *MYC* genes are both amplified. Generally, tumors that contain DMs or HSRs are in the advanced stages and require aggressive treatments. The molecular basis for the presence of DMs and HSRs in tumor cells is still not clear, but this type of chromosome abnormality is generally found in cells that have a mutation in the tumor suppressor gene that codes for p53 (see Chapter 16). p53 plays a crucial role in recognizing DNA damage and is central in stopping the cell cycle to allow repair of the DNA or directing a cell to apoptosis if the DNA cannot be repaired. One hypothesis suggests that chromosome abnormalities such as DMs or HSRs appear in a cell when it is damaged or when it ages. A normal cell with an active p53 will recognize these abnormalities and direct that cell to apoptosis. In the absence of p53, these chromosome abnormalities persist, and if they give a cell a growth advantage, that cell will continue to grow and develop into a tumor.

For more information see Hoffee PA: *IMS: Medical Molecular Genetics*. Fence Creek, pp. 219–221.

60 Prenatal Diagnosis

A Techniques for prenatal diagnosis

	CVS	*Amniocentesis*	*PUBS*
Gestational age performed	10–12 weeks	16–18 weeks	Greater than 18 weeks
Risk of miscarriage	0.5%	0.5%	1%
Benefits of screening	Early diagnosis	Relative safety	Provides blood sample for hematologic studies

B Prenatal diagnosis by amniocentesis

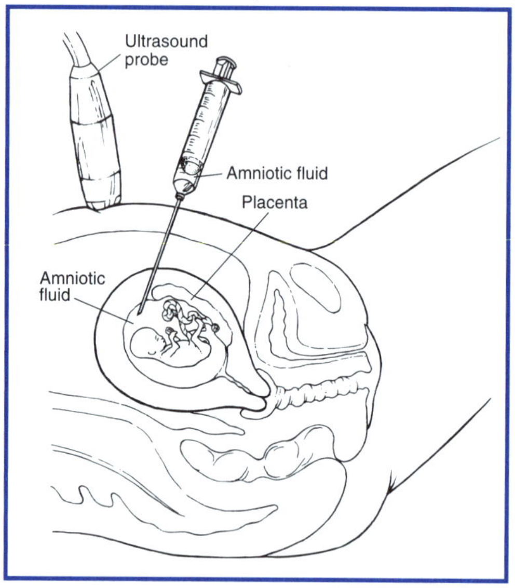

C Prenatal diagnosis by CVS

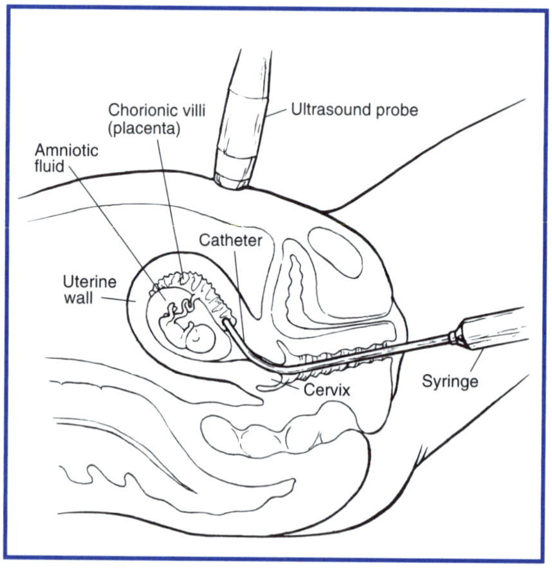

D Prenatal diagnosis by PUBS

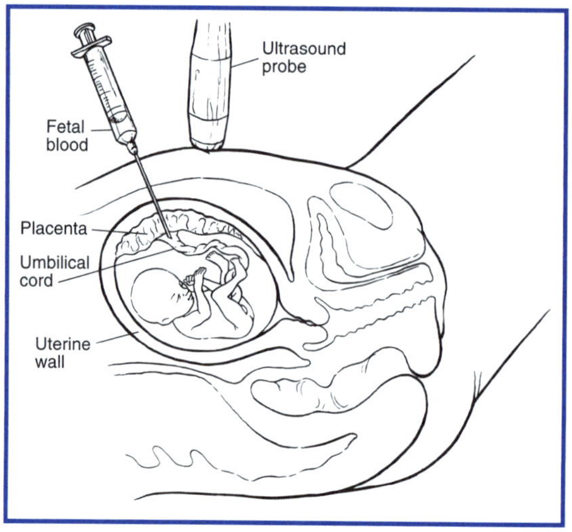

OVERVIEW

Birth defects currently account for a significant proportion of perinatal morbidity and mortality rates. However, during the past 20 years improved methods for prenatal diagnosis have made it possible for parents to alter the genetic risks for their children. Prenatal diagnosis allows monitoring of pregnancies at risk and provides three approaches to prevent genetic diseases: (1) screening newborns for disorders that can be treated early enough to prevent a clinical phenotype from arising, (2) testing parents who might be potential carriers of a genetic disease followed by genetic counseling to explain risk factors prior to their first pregnancy, and (3) diagnosing affected fetuses early in pregnancies to allow for genetic counseling of parents and other relatives at risk for the disease.

Indications for Prenatal Diagnosis

Prenatal diagnosis provides a method for determining whether a fetus at risk for a genetic disease is or is not affected with the disease. Not every pregnancy is assessed with a prenatal diagnosis because of the small but potential risk for miscarriage. Only pregnancies at high risk for genetic disorders are candidates for prenatal diagnosis, but not all at-risk pregnancies can be identified unless a couple has already had a child with a birth defect. There are, however, some common indicators for having prenatal diagnosis:

1. The mother is 35 years or older. (This is especially important in the detection of Down syndrome.)
2. The birth of a previous child who has trisomy 13, 18, or 21.
3. A previous child with a neural tube defect such as spina bifida or anencephaly, or spina bifida in either parent.
4. A biochemical or metabolic disorder is present in a previous child, or the parents are in a high-risk group for a particular inherited disorder such as Tay-Sachs disease or CF.
5. Demonstration of a chromosome abnormality, such as a translocation, in either parent.
6. Detection of abnormal maternal serum biochemical markers (AFP, uE_3, or HCG).

The three techniques used for prenatal diagnosis are summarized in Figure **A** and discussed below.

Amniocentesis

Amniocentesis is the aspiration of 20–30 mL of amniotic fluid from the amniotic cavity and is performed at 16–18 weeks gestation. Ultrasonography is performed before amniocentesis to determine fetal age, viability, number, anatomy, and placental location. Ultrasonography is used during the procedure for proper placement of a 22-gauge needle into the amniotic cavity, from which the fluid is removed (Figure **B**). The cells present in the fluid are usually cultured for cytogenetic analysis, which requires 2–3 weeks to obtain results. Direct analysis of the amniotic fluid proteins or extraction of DNA for the PCR can be carried out without cell culturing, and thus a more rapid diagnosis is possible when these techniques are used. The risks to the mother in this procedure are minimal; infection is the only rare complication. Fetal injury is usually not a problem with the continuous use of ultrasonography during the procedure. There is an increased risk (1 in 200) of spontaneous abortion. Amniocentesis is useful for the detection of chromosomal abnormalities, metabolic defects, and diseases that can be diagnosed by molecular techniques such as PCR. This diagnostic test is useful when a particular inherited disorder, such as the fragile X syndrome or Down syndrome, is suspected.

Chorionic Villus Sampling (CVS)

CVS is a biopsy of the placenta and is carried out at 10–12 weeks gestation. A small catheter placed through the cervix or a needle placed through the maternal abdomen allows the aspiration of chorionic villi that are loosely connected (Figure **C**). The villi have the same origin as the fetus and provide the same cytogenetic and biochemical information as the cells from the amniotic fluid. Ultrasonography is used before and during the procedure for proper placement of the catheter or needle into the developing placenta. For pregnancies at risk for neural tube defects, amniocentesis rather than CVS is performed. The risk to the mother is negligible, as with amniocentesis, and the risk of spontaneous abortion is the same, at 0.5%. CVS has two advantages over amniocentesis. It can be performed much earlier (10–12 weeks as opposed to 16–18 weeks) and can provide the information sooner to the parents. Additionally, the amount of material obtained by CVS is larger than with amniocentesis, allowing for direct assays on the uncultured cells. A disadvantage of the CVS procedure is a 1% failure rate in obtaining a sample. In determining whether a woman is a candidate for CVS, the risk of spontaneous abortion must be weighed against the risk of a child with a birth defect. It is important, therefore, to have a complete family history and an understanding of the chromosomal abnormality suspected so that the recurrence risk for the birth of an affected child can be evaluated (see Chapter 40).

Percutaneous Umbilical Blood Sampling (PUBS)

PUBS is used to obtain a sample of fetal blood and is performed after 18 weeks gestation. Continuous ultrasonography is used to guide a needle into the umbilical vein at either the placental or the fetal end of the umbilical cord (Figure **D**). Chromosomal analysis can be performed on samples obtained by PUBS within 48–72 hours and is useful if a rapid cytogenetic diagnosis is required. PUBS can be used to detect fetal infections or fetal blood disorders as well as to transfuse blood components or medication, or even to transplant fetal bone marrow. The risk of miscarriage or fetal death in this procedure is approximately 1%, and PUBS should be used only if the diagnosis cannot be made by amniocentesis or CVS or if a rapid diagnosis is required.

For more information see Hoffee PA: *IMS: Medical Molecular Genetics.* Fence Creek, pp. 315–321.

61 Gene Therapy I

Nonviral Vectors for Gene Delivery

A In vivo and ex vivo approaches to gene therapy

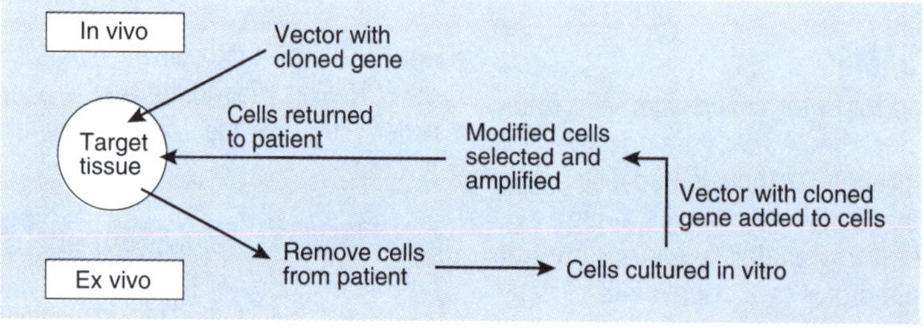

B Vector systems used for gene transfer

Vector system	*Advantages*	*Disadvantages*
Chemical and physical vectors		
Calcium phosphate	Easy to use	Random integration; inefficient DNA transfer
Electroporation	Easy to use; effective with many cell types; effective with embryonic stem cells	Random integration unless targeted; inefficient DNA transfer
Cationic liposomes	Noninfectious; nonimmunogenic; effective for in vivo transfer; can carry large DNA fragments	Unstable; remain episomal; poor gene expression
Polylysine-DNA conjugates	Targeted delivery; can carry large DNA fragments	Unstable; remain episomal; poor gene expression
Biologic (viral) vectors		
Retrovirus	Stable expression; broad host range; efficient transduction	Random integration; infects only dividing cells; shows variable expression; limit to size of DNA transferred
Adenovirus	Receptor-mediated uptake; infects nondividing cells; broad host range	Transient expression; remains episomal; elicits immune response; limit to size of DNA transferred
Adeno-associated virus	Potential site-specific integration; noninfectious; infects nondividing cells; broad host range	Difficult to produce; limit to size of DNA transferred; in developmental stages

C Cationic liposomes: a nonviral vector

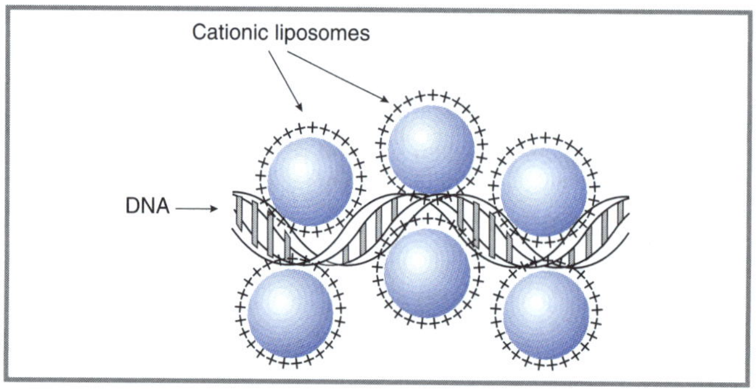

OVERVIEW

Gene therapy potentially can be used to treat inherited disorders in which a single gene product is missing or abnormal, as well as to treat acquired diseases such as cancers or infectious diseases. In all cases, a cloned gene must be transferred into the patient's cells, and once inside, it must be properly expressed in sufficient quantities to alleviate the clinical symptoms of the disease. Also, it is crucial to target the cloned gene to the tissue or organ that expresses the defect.

In Vivo and Ex Vivo Approaches to Gene Therapy

Two major approaches are used in gene therapy protocols to introduce cloned genes into cells: ex vivo methods and in vivo methods Figure **A**. In the ex vivo approach, cells are removed from the patient, grown in culture, genetically altered in vitro, and then returned to the patient. The in vivo approach involves the introduction of a gene directly into the patient's cells. The in vivo approach is preferred for gene therapy because it is theoretically more efficient and less expensive. Unfortunately, this approach is limited by the inefficiency in targeting genes to specific tissues and the inability to control which cells take up the transferred genes.

Gene Delivery Systems

Both in vivo and ex vivo gene therapy require an efficient method to introduce the gene into the cells as well as a suitable level of expression of the introduced gene. All the numerous ways developed to achieve gene transfer to mammalian cells have advantages and disadvantages, and no method is ideal. The methods used to transfer genes into cells can be classified into two major groups: chemical and physical methods and biological methods (Figure **B**). Once a gene or DNA enters a cell, it can have several fates. In some cases, portions of the added DNA are degraded in lysosomes, in which case this DNA never reaches the nucleus and is not expressed in the cell. Alternatively, the DNA can enter the nucleus undegraded, undergo recombination with the chromosomes of the cell, and become stably integrated into the genome. When integration of the exogenous DNA occurs, it most often does so by a process called nonhomologous recombination, which results in a random insertion of the DNA into the chromosomes at various sites. Only rarely does the exogenous DNA integrate into the chromosome at the site of its homologous gene. Even if the added DNA is not integrated at its normal site, it is associated with a chromosome and thus is stably expressed and transmitted to all daughter cells at mitosis. In some methods of gene transfer, the exogenous gene is not integrated into a chromosome but remains extrachromosomal in an episomal state. Because the episomal DNA does not contain a centromere, stable transmission to daughter cells does not occur. In the absence of any selective pressure, the genetic information is only transiently expressed, because the episomal DNA is lost after several cell divisions.

Nonviral Vectors

The transfer of DNA relies on an element called a vector that helps deliver the gene of interest to the cells. Chemical and physical elements used in transferring genes into cells are referred to as nonviral vectors. Two methods, calcium phosphate precipitation and electroporation, are generally not used for gene therapy studies; they are mainly used in vitro for preparing certain viral vectors. Calcium phosphate precipitation involves purifying DNA or cloning a desired gene, mixing the DNA with calcium and phosphate to produce a flocculating precipitate and allowing the mixture to precipitate on the surface of the cells. The cells take up the DNA by phagocytoses and deliver it into lysosomes. When the DNA is released from the lysosomes into the cytoplasm, it traverses the cytoplasm to the nucleus, where it crosses the nuclear membrane. Once inside the nucleus, the DNA can remain episomal to give transient expression, or a small portion may integrate randomly into a chromosome to give stable expression. The technique of electroporation involves exposing cells in the presence of DNA or a cloned gene to an electric field. The intensity of the electric charge causes localized pore formation in the plasma membrane, which allows the DNA to be taken up by the cells into endosomes. When the DNA exits the endosomes, it traverses the cytoplasm and enters the nucleus, where it is randomly integrated into a chromosome and is expressed. A cationic liposome is made up of a cationic lipophile and a neutral lipid. When mixed together, these two chemicals form a small unilamellar liposome that has the capacity to complex with DNA and create a lipid bilayer, with the DNA sandwiched in between the bilayers (Figure **C**). The positive charge of the DNA-liposome complex facilitates binding of the complex to the negatively charged surface of mammalian cells, allowing it to be taken up by endocytosis into endosomes. Once released from the endosomes into the cytoplasm, the DNA enters the nucleus, where it remains extrachromosomal. In this episomal state, the DNA is transiently expressed for only 7–10 days before it is lost from the cells. Cationic liposomes have several advantages as gene therapy vectors. They are capable of transferring large pieces of DNA, they are not infectious, and they do not elicit an immune response in the patient. The disadvantages of using cationic liposomes for gene transfer are their inefficient delivery to cells, the transient expression of the transferred genes, and the inability of the complexed DNA to be expressed at high levels. In spite of their limitations, cationic liposomes are potentially very useful as nonviral vectors for gene therapy and are currently being used in human clinical trials. Polylysine-DNA conjugates are complexes made up of DNA coupled to polylysine and allow the exogenous DNA to be targeted in vivo to specific tissues or organs. In this procedure, the DNA is hooked to a specific antibody by a polylysine chain. The presence of the antibody facilitates the interaction of the DNA complex with other proteins such as receptor ligands or viral proteins (see Chapter 62).

For more information see Hoffee PA: *IMS: Medical Molecular Genetics*. Fence Creek, pp. 333–337.

62 Gene Therapy II

Retroviral Vectors for Gene Therapy

A Life cycle of a retrovirus

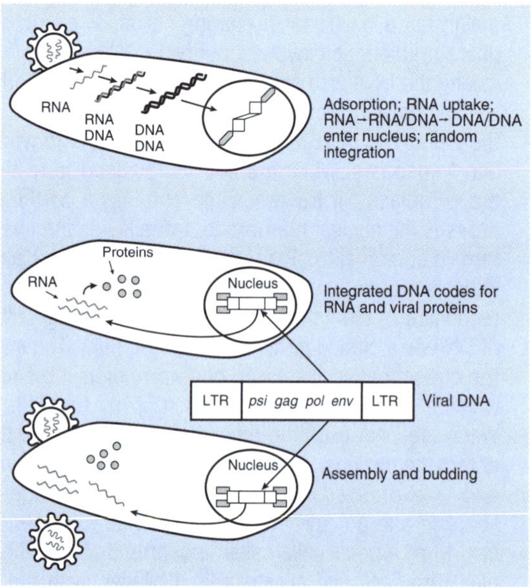

B Producing retroviral vectors for gene therapy

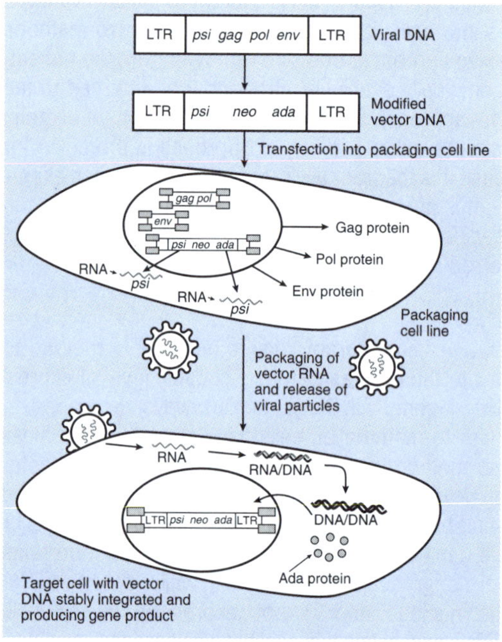

C Advantages and disadvantages of retroviral vectors

Advantages

1. They have the capacity to integrate a single copy of their genome into the host chromosomes. Even though the integration is random, it provides a means for creating a cell which stably maintains the transferred gene allowing for long-term expression of the gene.

2. The virus has a broad host range and gives efficient transduction with many kinds of growing cells, making it useful for many types of ex vivo gene therapy approaches.

3. The molecular biology of retroviruses is well developed.

DIsadvantages

1. Retroviruses are known to cause cancer in mammals. Because the potential, although slight, remains that replication-proficient retroviruses might be generated during the preparation process, vector preparations must be thoroughly tested before they are used in humans.

2. The viral vectors are capable of carrying only 8–9 kilobases of DNA, thus limiting the transfer process to smaller genes.

3. Retroviruses infect only actively growing cells, making them unsuitable for transferring genes to nongrowing or differentiated cells.

4. The random integration of the viral DNA into the host chromosomes possesses a small but potential chance for insertional mutagenesis, if the integration occurs within a crucial cellular gene.

OVERVIEW

The most common biological vectors currently in use in gene therapy protocols are mammalian virus derived vectors, which are genetically altered in such a way that they can still enter cells and express genes but can no longer replicate and become infectious. Both RNA- and DNA-containing viruses are used to develop vectors for gene therapy.

Retrovirus Vectors

The most frequently used and currently the best developed viral vectors for gene transfer studies are the retroviral vectors. Retroviruses are RNA-containing viruses that efficiently infect many types of growing cells and randomly insert a copy of their genome into the host chromosomes, producing a cell with a stably integrated viral genome. The life cycle of a retrovirus is shown in Figure **A**. The genome of a retrovirus is made up of RNA and contains information for three viral genes known as *gag, pol,* and *env*. The *gag* gene codes for the core protein of the virus, the *pol* gene codes for a reverse transcriptase needed to form a DNA intermediate from RNA, and the *env* gene codes for the envelope proteins of the virus. These genes, which code for *trans*-acting proteins, are required for the replication and production of the virus. In addition to the three genes that code for *trans*-acting proteins, the virus genome contains several important nucleotide sequences that act in *cis* and do not code for proteins. They include the 5′ and 3′ long terminal repeats (LTRs) that contain the promoter, poly-adenylation, and integration signals and a packaging sequence known as *psi*. The virus infects growing cells by binding to a specific receptor on the surface of cells that interacts with the viral envelope protein and results in a fusion of the virus and the cell membrane. Following this fusion, the virus undergoes endocytosis into endosomes. Release from the endosomes is accomplished by a viral protein–directed action. Once released from the endosome, the RNA genome is reversibly transcribed into DNA by the reverse transcriptase, resulting in a double-stranded DNA molecule that enters the nucleus and integrates randomly into the chromosomes of the cell. Entry into the nucleus appears to require breakdown of the nuclear membrane, a process that occurs only in mitosis, thus limiting retrovirus infection to cells that are undergoing cell division and active growth. Once the DNA of the virus has been integrated into the chromosomal DNA, it remains stable and is transcribed into RNA molecules that contain the entire viral gene sequences and are packaged into new viral particles. The presence of the viral DNA in the cell's chromosome does not affect the growth or metabolism of the cell, which allows stable cells that continually produce virus particles to survive and proliferate.

Producing Retroviral Vectors for Gene Therapy

The proteins coded for by the *gag, pol,* and *env* genes can be provided in *trans* and do not need to be present on the virus genome for replication and packaging of the RNA. Cell lines, called packaging cell lines, can be created that maintain these genes integrated into their chromosomes and provide the function of these genes to defective incoming retroviruses. However, the ability to package an RNA retroviral genome into a viral coat requires that the *psi* sequence be physically present on that RNA molecule. In creating a retroviral vector, the *gag, pol,* and *env* genes can be removed from the viral RNA without affecting the packaging of the RNA into new virions, provided that the viral LTRs, the packaging sequence, and a few other sequences important for reverse transcription are left intact. In a gene therapy protocol (Figure **B**), a modified viral DNA is created in vitro in which the *gag, pol,* and *env* gene sequences are replaced with the adenosine deaminase (*ada*) gene and the gene *neo*, which is a selective marker gene that makes a cell resistant to an antibiotic analogue called G418. The vector DNA retains the LTRs and the *psi* sequence. The modified viral DNA is transfected into a packaging cell line using calcium phosphate precipitation or electroporation, and the cells are plated in a medium containing G418. Cells that have integrated the defective vector DNA now contain the *neo* gene and thus are stably resistant to G418. The cell lines with the modified viral vector stably integrated into a chromosome are the producer cell lines, which have the capacity to produce retroviral vectors that carry the *ada* gene. The cell lines contain the viral genes provide the Gag, Env, and Pol proteins in *trans*, but the RNA molecules that produced that code for these functions do not carry a *psi* sequence. The defective viral vector DNA produces a RNA molecule containing the LTR, the *psi* sequence, the *ada* gene, and the *neo* gene. In the presence of the *trans*-acting proteins, only the RNA molecules containing the *psi* sequence, and thus the *ada* gene and the *neo* gene, will be encapsulated and packaged into virus particles, which now bud from the cell into the medium. The viral particles released can be directly added to a target cell or first purified and then added to target cells. These mature virus particles, even though defective, can still infect a target cell by the normal procedure, but only once. Because they carry a functional reverse transcriptase within the viral particle, the defective viral vector will make a DNA copy of its RNA and stably integrate the DNA into the chromosomes of the target cell. In this way, the target cell now contains a stably integrated and functional gene, such as the *ada* gene. The absence of any viral genes in the target cell and on the incoming defective viral vector genome prevents any further production of virus particles from the target cell.

Advantages and Disadvantages of Retroviral Vectors

The use of retroviral vectors for gene therapy clinical trials has some advantages as well as disadvantages (Figure **C**). All factors must be taken into account when these vectors are used in a clinical trial.

For more information see Hoffee PA: *IMS: Medical Molecular Genetics*. Fence Creek, pp. 338–341.

63 Gene Therapy III

Adenoviruses, Adeno-Associated Viruses, and Mammalian Artificial Chromosomes

A DNA insert size for different vectors

Vector	Insert size
Adeno-associated virus	4.5 kb
Adenovirus (1)	7–8 kb
Retrovirus	8–9 kb
Adenovirus (2)	28 kb
Cationic liposome	> 50 kb
Human artificial chromosome	> 50 kb

B Adenovirus-polylysine-DNA complexes for gene transfer

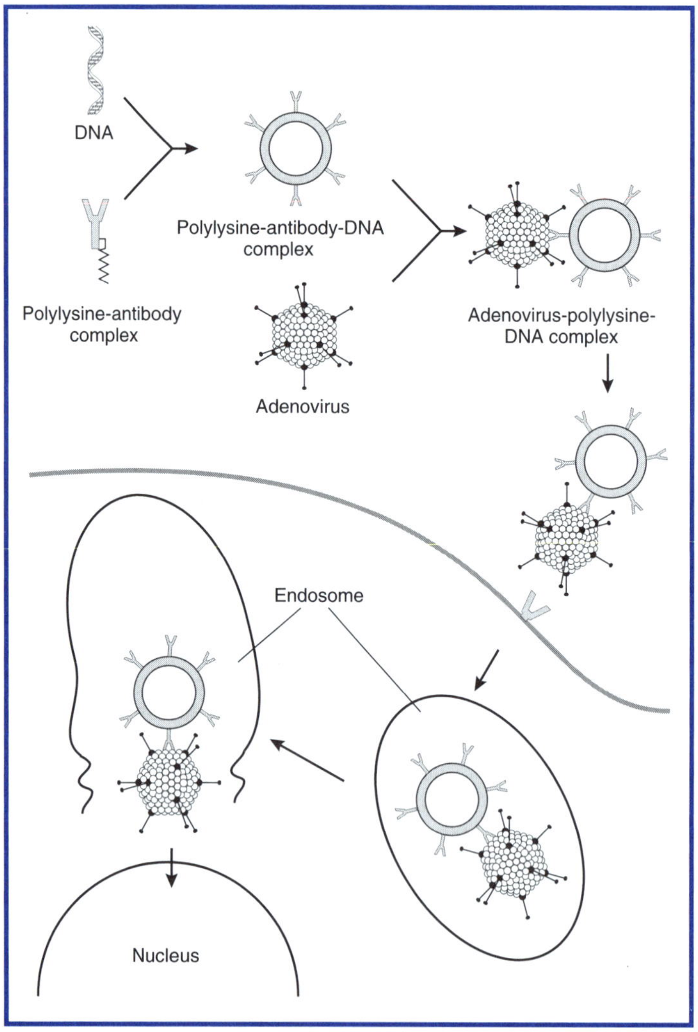

C Elements of a human artificial chromosome

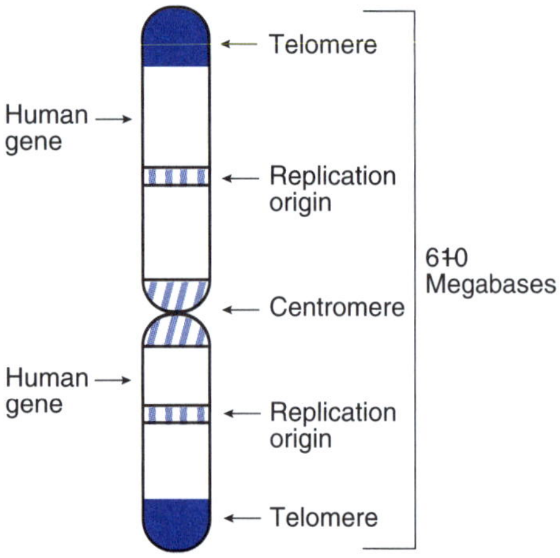

D Advantages of mammalian artificial chromosomes as a gene therapy vector

1. A human artificial chromosome can accommodate large human DNA inserts, removing the limitation on the size of the gene to be transferred.

2. The presence of a centromere sequence within the chromosome assures its stability and its equal distribution to daughter cells at mitosis.

3. The addition of telomere sequences to the chromosome ends provides stability for the chromosome structure and prevents its degradation and fusion with other normal chromosomes present in the nucleus.

4. The presence of DNA replication origins on the chromosome maintains its normal copy number from one cell cycle to the next by regulating its replication during the normal S phase of the cell cycle.

OVERVIEW

In their search for the ideal vector for gene therapy, scientists are actively testing other viral vectors as well as nonviral vectors as a means of delivering genes to human cells and tissues. Vectors are needed that deliver genes with high efficiency into the proper tissue and, once there, stably express the transduced gene at the right time and at the appropriate level. Additionally, the ideal vector should either remain in a stable extrachromosomal state or have the ability to target a specific site within the genome, where it can undergo homologous recombination with the host cell DNA.

Adenovirus Vectors as Vectors for Gene Therapy

Adenoviruses are DNA-containing viruses that normally produce respiratory infections in humans. The adenoviral genome is 38 kb in size and is modified for gene therapy protocols by the deletion of two viral genes, leaving approximately 7–8 kb of space for insertion of the gene of interest (Figure **A**). Adenoviruses have a broad host range and infect both dividing and nondividing cells. They enter the cell by receptor-mediated endocytosis and have an effective mechanism to disrupt endosomes, making them potentially ideal virus vectors for gene therapy. On the negative side, once the adenoviral DNA is inside the nucleus, it does not integrate into the chromosomes but remains episomal, making stability and long-term expression of transferred genes a serious problem. The results from initial experiments using the deleted adenoviral vectors for gene therapy have been disappointing because of inflammation and neutralizing antibodies against the viral proteins developing in the patients (see Chapter 64). This adverse host reaction limits repeated administration of the virus, which is necessary because of the lack of stable long-term gene expression that occurs with the episomal adenoviral DNA. Recently, two new types of adenoviral vectors have been developed. One is an adenovirus vector in which all of the protein-coding viral genes have been deleted. This allows the vector to accommodate larger DNA inserts, up to 28 kb, and to avoid the production of any viral proteins that might elicit an immune response. This new adenoviral vector retains only the long terminal repeat sequences and the packaging sequences of the virus and requires a helper virus for its preparation. It is currently being developed for use in gene therapy for Duchenne muscular dystrophy (DMD). The second is an adenovirus-polylysine-DNA complex that utilizes only the adenoviral coat proteins that are coupled to DNA via antibodies and polylysine (Figure **B**). The antibody, which recognizes the adenovirus protein coat, facilitates the formation of a complex that has the DNA hooked through the polylysine chain to the antibody that is bound to the virus shell. When this adenovirus-polylysine-DNA complex is mixed with mammalian cells, the complex binds to specific receptors on the surface of the cell and is internalized by receptor-mediated endocytosis directly into endosomes. The adenovirus proteins serve two functions: to bind to the cell surface receptors and facilitate receptor-mediated endocytosis, and to use the capacity of the proteins to stimulate disruption of the endosomes, thus releasing the DNA more efficiently into the cytoplasm. A more efficient release of DNA from the endosomes, a limiting step in several methods of gene transfer, results in an increased amount of the DNA reaching the nucleus, thereby resulting in an increased expression of the transferred gene. Additionally, proteins can be included in the complex that recognize cell surface receptors other than the virus receptors. This modification helps target the complex to specific cells, even though these cells may not have the adenovirus receptor. Using these adenovirus-polylysine-DNA complexes for gene therapy is a promising approach, but this method of gene transfer has had limited success because of the transient retention of the DNA in the nucleus and the poor expression of the transferred gene. This new form of adenovirus vector may help overcome the problem of evoking a nonspecific inflammation and cellular immunity that has been associated with previous adenovirus vectors; yet, it retains the important properties of receptor-mediated endocytosis and the ability to facilitate release of the transferred genes from the endosomes.

Adeno-Associated Viruses as Vectors for Gene Therapy

Adeno-associated viruses are single-stranded DNA viruses that require the help of another virus, such as adenovirus, to carry out a productive infection. When a helper virus is not available, the adeno-associated viral DNA integrates into the chromosome. The integration process is highly specific, occurring on the q arm of chromosome 19 at 19q13, and results in a stable long-term persistence of the viral genome without any obvious harm to the host cells. The virus is small and can accommodate only 4.5 kb of DNA, limiting its usefulness for transferring larger genes. For smaller genes, however, adeno-associated virus remains an attractive candidate for gene therapy and is capable of transducing genes to a wide range of cell types, at least in vitro. At present the virus is difficult to produce in large quantities, and it is still in the developmental stages as a gene therapy viral vector.

Mammalian Artificial Chromosomes as Vectors for Gene Therapy

The creation of a mini-artificial human chromosome vector containing human DNA sequences, a centromere sequence, telomere sequences, and DNA replication origins offers several advantages for gene therapy that many other vectors lack (Figures **C** and **D**). These nonviral biological vectors are currently under development, and they offer a new type of vector for future gene therapy studies.

For more information see Hoffee PA: *IMS: Medical Molecular Genetics*. Fence Creek, pp. 341–342.

64 Gene Therapy and the Treatment of Human Diseases

A Examples of human diseases amenable to gene therapy

Disease	*Cells treated*	*Approach used*
Inherited diseases		
Adenosine deaminase (ADA) deficiency	Stem cells	Ex vivo *ada* gene addition
Gaucher disease	Stem cells	Ex vivo glucocerebrosidase gene addition
Cystic fibrosis	Respiratory epithelium	In vivo *CFTR* gene addition
Familial hypercholesterolemia	Liver cells	Ex vivo LDL receptor gene addition
Duchenne muscular dystrophy	Muscle cells	In vivo dystrophin gene addition
Acquired diseases		
Malignant melanoma	Fibroblasts	Ex vivo interleukin-2 gene addition
Brain tumors	Tumor cells	In vivo addition of herpes simplex virus thymidine kinase gene
Ovarian cancer	Tumor cells	In vivo addition of p53 gene
Neck cancer	Tumor cells	In vivo infection with EB1 defective adenovirus

Note: CFTR = Cystic fibrosis transmembrane conductance regulator

B Gene therapy for treating arthritis

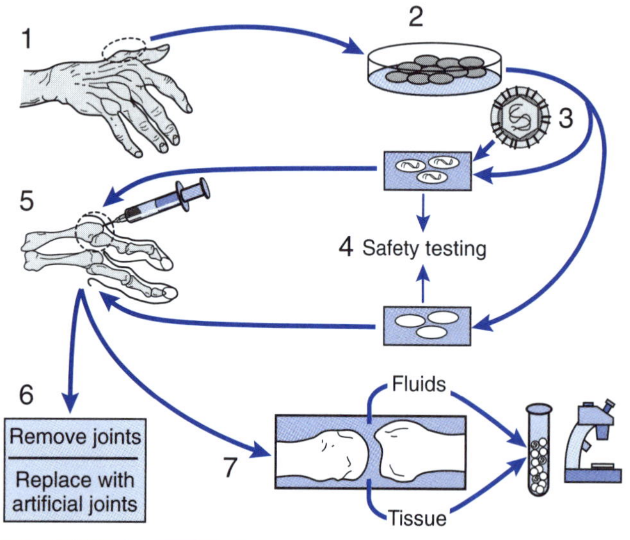

1. Synovial tissue removed from joint of patient.
2. Synovial cells cultured for 1 week.
3. Cell culture split in two; 1/2 treated with retrovirus with interleukin-1 receptor antagonist gene; 1/2 not treated (2 weeks)
4. Virus-exposed cells tested for expression of gene. Cells from both cultures tested by FDA-approved facility for safety (6 to 7 weeks). Remaining cells frozen.
5. Once declared safe, cells are thawed and injected into patient's arthritic knuckle; two knuckles are injected with transfected cells and two injected with untreated cells.
6. Remove joints and replace with artificial joints.
7. Tissue and fluid from the removed joints are analyzed for any reduction in inflammation and for expression of the transfected gene.

OVERVIEW

Several different strategies are currently available for using gene therapy to treat both inherited and acquired human diseases. In dealing with inherited human diseases that are caused by the loss of function of a single gene, the gene addition approach, which involves the introduction of copies of a functional gene into the appropriate cell, is most often used. In these experiments it is difficult to control the level of expression within the proper cell type, and it is impossible to control the site of integration of the added gene when retroviral vectors are used. Despite these limitations, if a low level of enzyme is sufficient to reverse the disease phenotype, then a gene addition procedure can theoretically be successful for treating inherited diseases such as those listed in Figure **A**.

Examples of Inherited and Acquired Human Diseases Amenable to Gene Therapy

The best known example of a successful gene addition approach to correct a disease phenotype is the treatment of a form of severe combined immunodeficiency disease (SCID) that results from the loss of adenosine deaminase (ADA) activity. The absence of ADA activity leads to the loss of both B- and T-cell function and is usually lethal, with patients dying of infections or cancer within the first few years of life. The treatments available for the disease include bone marrow transplantation, if a suitable donor is available, and enzyme replacement therapy using polyethylene glycol–conjugated ADA protein. In 1990, the first ADA-deficient patient was treated by gene therapy using an ex vivo method and a retroviral vector system. The normal *ada* gene of 1.5 kb was cloned into a retroviral vector (see Chapter 62), and the genetically modified retroviral vector was transfected into T lymphocytes that had been removed from the patient and cultured in vitro. The modified T cells, which contained a retroviral vector carrying a normal *ada* gene integrated into their chromosomal DNA, were then reinfused into the patient. After 5–6 months of gene therapy treatment, which involved repeated infusions of the retroviral-transfected T cells, the peripheral blood T-cell count of the patient had increased and stabilized, and her ADA activity had increased from nondetectable levels to one-half the level seen in an individual heterozygous for the *ada* gene. After 2 years of treatment, the level of T cells and the levels of ADA activity remained stable even when the gene therapy treatment had been discontinued. After 5 years, her T-cell count and her level of ADA activity still remained stable, although no more than 50% of her T cells contained the normal *ada* gene. The apparent success of the *ada* gene therapy clinical trials launched a whole series of additional clinical trials that attempted to use the gene addition approach to correct a variety of different inherited human diseases, such as CF, familial hypercholesterolemia (FH), and DMD. These inherited diseases have common features: they are associated with single gene defects that have been identified at the molecular level and have a well-characterized pathology; all of them are fatal and usually affect children; and most conventional therapies have proved inadequate in their treatment. At least in animal models, some success has been achieved when a normal gene is provided to cells and correction of the clinical phenotype is observed. Human clinical trials have been less successful, and it is unclear whether any firmly documented clinical benefits have occurred. Attempts to treat CF with the CFTR gene using adenoviral vectors failed to alleviate the ion transport deficiency in the disease. Failure was attributed to the low efficiency of gene transfer to cells in vivo and the inflammatory and immune response to the viral vector. The attempts to treat patients with FH by partial hepatectomy, transfection of hepatocytes with the gene for the LDL receptor in culture, and reinfusion of the modified cells into the portal vein of patients has likewise met with less than impressive results. The use of myoblasts to treat DMD has also been unsuccessful, although more recent approaches using completely defective adenoviral vectors carrying a mini-dystrophin gene offer promise. On the positive side, much has been learned from the less than successful trials concerning methods of transferring genes into human cells, how to achieve continued and efficient expression of the transferred genes in vivo, and identification of the problems that must be met in dealing with a complicated organism like the human body rather than cells in culture. Most gene therapy clinical trials in progress are designed to treat human cancer, an acquired disease. There are several approaches to treating acquired diseases with gene therapy (see Figure **A**). In one approach, tumor cells or fibroblasts are genetically modified to produce compounds that stimulate the normal host immune system. In this way, the enhanced immune system attacks and kills cancer cells. In a second approach, the cancer cells are directly killed by the production of toxic compounds that specifically attack the cancer cells. A third approach utilizes compounds such as anti-sense RNAs that block the expression of oncogenes in cancer cells. Finally, if the patient is known to have a defective tumor suppressor gene, such as the *p53* or the *RB* gene, theoretically one can add that gene back to the patient's cells.

Gene Therapy for Arthritis

Rheumatoid arthritis, an autoimmune disease, causes chronic and painful inflammation in the joints of affected individuals. The disease has no known cure but is a potential candidate for treatment with gene therapy. The steps outlined in Figure **B** illustrate a current protocol being tested to treat arthritis by gene therapy using a retrovirus vector. In this protocol, a retrovirus vector containing a gene for the interleukin-1 receptor antagonist protein is transferred to synovial cells ex vivo, and the cells are then reintroduced into the damaged joints in an attempt to obtain a concentrated accumulation of the therapeutic protein at the site of the disease. The procedure involves several time-consuming steps, including culturing and thorough analysis of cells as well as extensive testing for safety of the preparations, before they are used to treat patients.

For more information see Hoffee PA: *IMS: Medical Molecular Genetics*. Fence Creek, pp. 343–348.

65 Animal Models of Human Disease

A Generation of transgenic mice from embryo stem cells

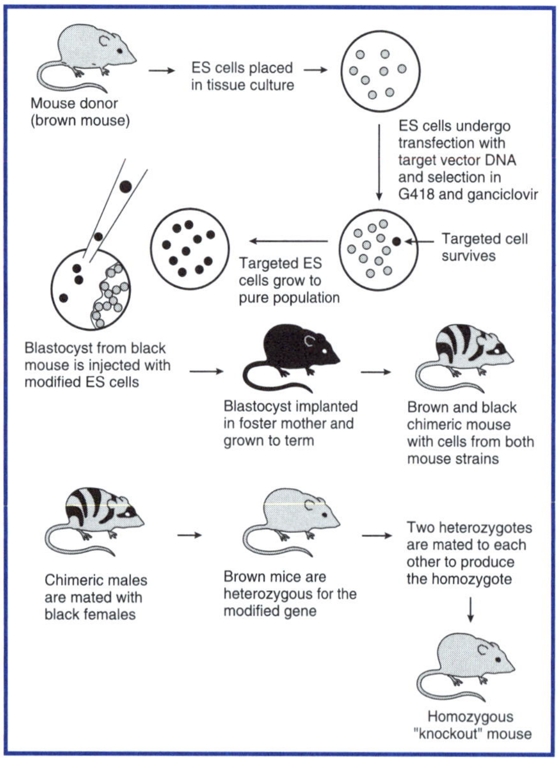

C Cloning of mammals from adult DNA

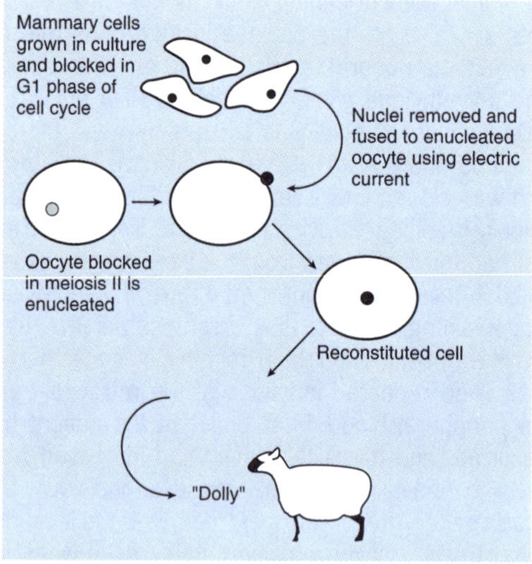

B Examples of transgenic mouse models

Human disease	*Protein altered*	*Mouse gene*
Osteogenisis imperfecta	Collagen, alpha 1	*Cola 1*
Achondroplasia	Fibroplast growth factor receptor 3	*Fgfr3*
Alzheimer disease	Amyloid beta precursor	*App*
Ataxia telangiectasia	Phosphoinositol kinase	*Atm*
Myotonic dystrophy	DM kinase	*DM15*
Duchenne muscular dystrophy	Dystrophin	*Dma (mdx)*
Fragile X syndrome	FMR1	*Fmr 1*
Huntington disease	Huntingtin	*Hdh*
Acute intermittent porphyria	Porphobilinogen deaminase	*Pbgd*
Adematous polyposis coli	APC	*Apc*
Breast cancer, type 1	BRCA1	*Brca1*
Malignant melanoma	Cyclin-dependent kinase inhibitor p16	*Cdkn2A*
Hereditary nonpolyposis colorectal cancer	hMSH2, hMLF1, hPMS2	*Msh2, Mlh1, Pms2*
Neurofibromatosis 1	Guanosine triphosphatase	*Nf1*
Retinoblastoma	Rb protein	*Rb1*
Li-Fraumeni syndrome	p53	*Trp53*
Severe combined immunodeficiency syndrome	Adenosine deaminase	*Ada*
α-Thalassemia	Hemoglobin alpha	*Hba*
β-Thalassemia	Hemoglobin beta	*Hbb*
Cystic fibrosis	CFTR	*Cftr*
Gaucher disease, type I	Glucocerebrosidase	*Gba*
Tay-Sachs disease	Hexokinase A	*Hexa*
Lesch-Nyhan syndrome	Hypoxanthine-guanine phosphoribosyltransferase	*Hgprt*
Familial hypercholesterolemia	Low-density lipoprotein receptor	*Ldlr*
Phenylketonuria	Phenylalinine hydroxylase	*Pah*

OVERVIEW

A transgenic mouse is a mouse with a modified gene stably integrated into the genome of its germ cells, allowing that gene to be passed from one generation to the next. For the most part, the transgenic mice that are used as model systems today are mice in which a specific gene has been altered or knocked out. These defective mice are used in explorations of the various effects that mutations have on development, immunity, and other metabolic processes.

Generation of Transgenic Mice from Embryonic Stem (ES) Cells

A common method developed to create transgenic mice involves inserting a single copy of a gene into a targeted site of the mouse genome by homologous recombination (Figure **A**). The specificity of this process allows the creation of transgenic mice that harbor a corresponding mutation for any human disease gene that has been cloned. This approach became possible with the development of a method to culture pluripotent ES cells in vitro and maintain their pluripotency even after multiple cell divisions. In this procedure, mouse ES cells are taken from the inner mass of a blastocyst of a congenic 129/SvJ strain of mice that has an agouti (brown) coat color. The ES cells are grown in culture and then transfected with a vector designed to promote recombination with homologous gene sequences in the mouse genome. The vector carries DNA sequences of the gene to be targeted, along with a positive gene marker and a negative gene marker that can be used to isolate ES cells that have integrated the vector DNA into a specific site. Once an appropriate ES cell has been isolated and shown to have a specific gene inactivated, the cell is amplified in culture to produce a population of genetically modified ES cells. The genetically modified ES cells are microinjected into a blastocyst derived from a mouse line that has a different coat color, such as the C57BL/6J mouse, or black coat mouse. The black coat color gene is recessive to the agouti coat color gene, allowing one to visualize chimeric mice that arise. The blastocyst is then implanted in a foster mother with a black coat, and the offspring are allowed to grow to term. Any newborn mice with a chimeric or mixed-coat color of brown and black are evidence that the modified ES cells have survived and proliferated in the mouse. Identification of chimeric mice that have the targeted mutation in the germ cells requires mating male chimeric mice to female black mice and looking for brown mice that are produced when the sperm is derived from the original genetically modified ES cells. The brown mice are then analyzed by molecular techniques to determine whether they carry a copy of the mutated gene at the proper genomic site. For genes other than those located on the X chromosome, only one allele of a gene is inactivated by the recombinational event that occurs between the vector DNA and the host chromosome. The selection of mice that have the second allele inactivated occurs during the subsequent breeding process. Matings between mice that are heterozygous for the mutated gene produce offspring that are homozygous for the mutated gene at a 25% frequency. These homozygous knockout mice now serve as model systems that can be studied for the phenotypic consequences of a mutation in a specific gene.

Examples of Transgenic Mouse Models

Many inherited human diseases have knockout mice available to study as model systems (Figure **B**). Mouse models are used to study the function of a disease gene in an otherwise normal genetic background, to understand the pathophysiology of the disease associated with a specific gene mutation, and to provide models to develop gene therapy protocols for alleviating the clinical phenotype of a disease. There are, however, problems associated with some of the model systems, since a mouse, after all, is not a human. Mice may have alternative biochemical and developmental pathways that sometimes influence the expression of the transgene. Furthermore, the targeting procedure used to develop the transgenic animals provides a mouse model in which the mutation produced is a loss-of-function mutation, a result of the endogenous gene function being destroyed by the transgene integration. Such models exclude the study of diseases that result from gain-of-function mutations, such as the human diseases that result from trinucleotide expansion within a gene, like myotonic dystrophy (a trinucleotide expansion in the 3′ untranslated region of the gene) and Huntington disease (a trinucleotide expansion in the 5′ translated region of the disease).

Cloning of Mammals from Adult DNA

Although species such as frogs, mice, cattle, and monkeys can be cloned by using the nuclei from embryonic cells as discussed above, no one was able to clone a mammal with DNA taken from adult cells until recently. The problem in using adult DNA to clone an animal appears to be related to the irreversible changes that occur in DNA during development. In an attempt to overcome this irreversibility, Wilmut and colleagues removed cells from mammary tissue and blocked the adult cells in the G1 phase of the cell cycle (Figure **C**). When the blocked mammary cells were fused with an enucleated egg cell, the adult DNA received a signal to bypass the G1/S border and enter into the S phase of the cell cycle, where the DNA begin to replicate using the cytoplasmic proteins provided by the egg cell rather than the old proteins previously attached to the adult DNA. The early lack of gene expression when the cells are blocked in G1 is believed to reprogram the DNA so that it is capable of undergoing a normal development from an embryo to an adult. The efficiency of this cloning process was very low, with only one reported success, the sheep Dolly, in 277 attempts. Using fetal cells as the DNA donor rather than adult cells in this procedure is much more efficient. Additionally, animals can be cloned that carry therapeutic genes, such as the human clotting factor IX, by transfecting these foreign genes into the fetal cells before they are used as nuclear donors.

For more information see Hoffee PA: *IMS: Medical Molecular Genetics*. Fence Creek, pp. 348–353.

PART IV: QUESTIONS

Directions: For each of the following questions, choose the **one best** answer.

Question 1

1. Which trisomy listed below is incompatible with life?

(A) Trisomy 1

(B) Trisomy 13

(C) Trisomy 18

(D) Trisomy 21

(E) Trisomy X

Directions: The groups of questions below consist of lettered choices followed by several numbered items. For each numbered item, select the appropriate lettered option with which it is most closely associated. Each lettered option may be used once, more than once, or not at all.

Questions 2–4

A recent newspaper article reported the case of Karen, a 680-lb, 13-year-old girl who died of heart failure. The girl's mother stated that Karen had an uncontrollable compulsion to eat and ate constantly. The mother did not deny her food because the girl would have violent outbursts and tantrums unless she got the food she craved. Karen also had numerous bedsores, at which she constantly picked.

2. The phenotype of this child is consistent with which one of the following syndromes?

(A) Down syndrome

(B) Angelman syndrome (AS)

(C) Prader-Willi syndrome (PWS)

(D) Patau syndrome

(E) Turner syndrome

A DNA sample was obtained from Karen at autopsy and from both parents and used for karyotype analysis and Southern blotting. The karyotypes of all three individuals (the mother, the father, and Karen) were normal. DNA probes for chromosomes 3, 9, 15, and 17 were used to analyze restriction length polymorphisms in the DNA of the mother (M), the father (F), and Karen (K) and gave the following results:

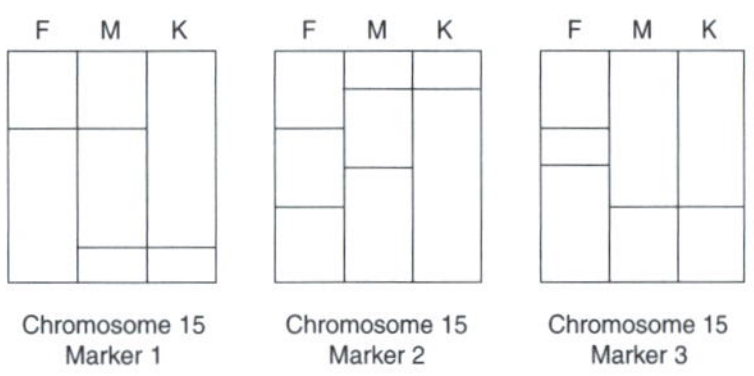

Several other markers on various chromosomes were also tested:

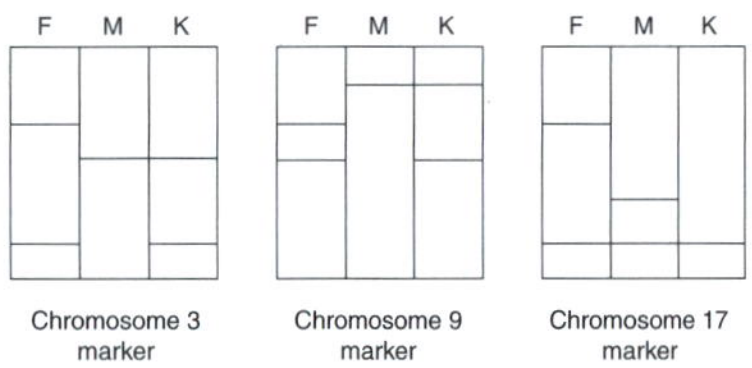

3. The results of this DNA analysis are most consistent with

(A) uniparental diploidy

(B) uniparental disomy for chromosome 15

(C) uniparental disomy for chromosome 9

(D) a deletion on chromosome 3

(E) a deletion on chromosome 17

4. A father is a carrier of a Robertsonian translocation. His risk of having an affected child is highest when he carries which one of the following translocations?

(A) der(14;21)(q10;q10)

(B) der(21;22)(q10;q10)

(C) der(21;21)(q10;q10)

(D) der(13;14)(q10;q10)

(E) der(14;15)(q10;q10)

Directions: For each of the following questions, choose the **one best** answer.

Questions 5–8

5. Inactivation of the X chromosome is controlled at the molecular level by which of the following mechanisms?

(A) Protein coded for by a gene on the X chromosome

(B) Protein coded for by a gene on an autosome

(C) RNA molecule coded for by a gene on the X chromosome

(D) RNA molecule coded for by a gene on an autosome

(E) Deletion of multiple genes on the inactive X chromosome

6. Burkitt's lymphoma is a result of which of the following molecular changes?

(A) Translocation that alters the structure of the Rb protein

(B) Translocation that alters the structure of the Abl protein

(C) Translocation that alters the expression of the *abl* gene

(D) Translocation that alters the expression of the *myc* gene

(E) Mutation that inactivates a tumor suppressor gene

7. Which one of the following nucleotide sequence is required in *cis* for a retroviral RNA to be packaged into a viral particle?

(A) *gag* nt sequence

(B) *env* nt sequence

(C) *pol* nt sequence

(D) *psi* nt sequence

(E) Promoter nt sequence

8. Which one of the following vectors used in gene therapy studies normally integrates into a specific site on a host chromosome?

(A) Retroviral vectors

(B) Adenoviral vectors

(C) Adeno-associated viral vectors

(D) Cationic liposomes

(E) Human artificial chromosomes

Directions: The groups of questions below consist of lettered choices followed by several numbered items. For each numbered item, select the appropriate lettered option with which it is most closely associated. Each lettered option may be used once, more than once, or not at all.

Questions 9–13
Match the following chromosomal abnormalities with the appropriate description.

(A) Reciprocal translocation

(B) Robertsonian translocation

(C) Interstitial deletion

(D) Paracentric inversion

(E) Isochromosome

9. There is a fusion between two acrocentric chromosomes.

10. There is a division along the wrong axis of a chromosome that results in a chromosome with two copies of one arm and no copy of the other arm.

11. A balanced exchange occurs between two chromosomes, and no genetic material is lost.

12. Two breaks occur on the same arm of a chromosome, and the DNA located between the two breaks is rejoined in an inverted order.

13. Genetic material is deleted from within a chromosome.

Questions 14–16
Match each of the following syndromes with the appropriate karyotype:

(A) Turner syndrome

(B) Edwards syndrome

(C) Androgen insensitivity syndrome

(D) Angelman syndrome

(E) Klinefelter syndrome

14. Male with a 47,XXY karyotype

15. Female with a 46,XY karyotype

16. Female with a 45,X karyotype

PART IV: ANSWERS AND EXPLANATIONS

1. **The answer is A.**

The only three autosomal trisomies compatible with life are trisomy 13 (Patau syndrome), trisomy 18 (Edwards syndrome), and trisomy 21 (Down syndrome). Trisomy X occurs in approximately 1 in 1000 liveborn females but is usually not associated with any phenotypic abnormalities.

2. **The answer is C.**

Individuals affected with PWS have an uncontrollable appetite, and if not kept in check they will gain enormous amounts of weight. Additionally, these individuals tend to have behavioral problems such as tantrums and are known to pick at sores. The only way to control the weight gain in these individuals is with strict dieting and constant exercise.

3. **The answer is B.**

The probes used recognize unique DNA segments on chromosomes 3, 9, 15, and 17. Normal individuals have DNA fragments that come from both the mother and the father. However, if a child has received both copies of one chromosome from either the mother or the father, only the maternal or paternal alleles, respectively, are present. The analysis of the three markers from chromosome 15 suggests that the child received only the maternal alleles and none of the paternal alleles. Thus, she must have received both number 15 chromosomes from the mother, a phenomenon known as uniparental disomy. The controls show that the child has received both maternal and paternal alleles for chromosomes 3, 9, and 17. Uniparental disomy for chromosome 15, in which the child receives both maternal chromosomes 15, is consistent with a diagnosis of PWS.

4. **The answer is C.**

Since the father carriers two copies of chromosome 21 as a single chromosome, all his children have a risk of receiving three copies of chromosome 21: one from the mother and two from the father. Therefore, all his children will be trisomy 21 and have Down syndrome. In the other cases, a normal chromosome is also present and it will be inherited by his offspring 50% of the time and they will be normal.

5. **The answer is C.**

The *XIST* gene located on the X chromosome codes for a RNA molecule that functions directly as a RNA molecule and does not code for a protein. XIST transcripts bind to the inactive X chromosome and act in *cis*. The stability of the XIST RNA appears to be the important factor in determining which X chromosome is to be inactivated.

6. **The answer is D.**

The chromosome translocation t(8;14)(q24;q32) is found in approximately 85% of patients with Burkitt's lymphoma. This translocation moves the *myc* gene from its normal location on chromosome 8 into the middle of the immunoglobulin heavy chain gene on chromosome 14 and places the *myc* gene under the control of a new promoter. This promoter increases the expression of the *myc* gene and results in overproduction of the *myc* gene product. A translocation that alters the Rb protein will give rise to RB. A translocation that alters the structure of the Abl protein gives rise to CML.

7. **The answer is D.**

The *psi* sequence, known as the packaging sequence, must be present on a retroviral RNA molecule for packaging of that RNA into viral particles to occur. The *gag, env,* and *pol* genes code for protein products that can be provided in *trans*. The promoter sequence is required for expression of the genes and not for packaging.

8. **The answer is C.**

Adeno-associated virus integrates on the q arm of chromosome 19 at 19q13 and persists in a stable long-term association with the host cell. Retroviruses also undergo stable integration into the host chromosomes, but the site of integration is random. Adenoviruses, cationic liposomes, and human artificial chromosomes remain in an extrachromosomal state and do not normally integrate into the host chromosomes.

9–13. **The answers are 9-B, 10-E, 11-A, 12-D, 13-C.**

A Robertsonian translocation is the result of the fusion of two acrocentric chromosomes (i.e., chromosomes 13, 14, 15, 21, 22) at or near the centromeres, with the loss of the short arms from both chromosomes. An isochromosome is the result of a division along the wrong axis of a chromosome, resulting in a chromosome with two copies of one arm and no copies of the other arm. An isochromosome is usually lethal except for those involving the X chromosome or chromosome 21. A reciprocal translocation is a balanced exchange between two chromosomes with no loss of genetic material. Individuals with a balanced translocation are phenotypically normal. Inversions occur when a chromosome breaks at two points and may involve the centromere (pericentric inversions) or may not involve the centromere (paracentric inversions). An interstitial deletion occurs when a chromosome breaks at two sites, and the material between the two breaks is deleted.

14–16. **The answers are 14-E, 15-C, 16-A.**

Individuals with Klinefelter syndrome have a 47,XXY karyotype and suffer from hypogonadism. Classic features of Klinefelter syndrome include tall stature, small testes, and infertility. Androgen insensitivity syndrome is a result of a mutation in the androgen receptor gene located on the X chromosome and gives rise to 46,XY females. Approximately 50% of individuals with Turner syndrome have a 45,X karyotype with the absence of the X chromosome, a result of nondisjunction in females or anaphase lag in males.

References

Chapter 1

1. Watson, J.D., Crick, F.H. Molecular structure of nucleic acid. A structure for deoxyribose nucleic acid. Nature, 171:737–738, 1953.

Chapter 2

1. Matthews, C., Van Holde, K., 2nd ed. Biochemistry, Menlo Park, CA, Benjamin/Cummings Publishing Co. 1996.

Chapter 3

1. Becker, W.M., Reece, J.B., Poenie, M.F. The World of the Cell, 3rd ed. Menlo Park, California. Benjamin/Cummings, 1996.
2. Holt, S.E., Shay, J.W., Wright, W.E. Refining the telomere-telomerase hypothesis of aging and cancer. Nature Biotechnology, 14:836–839. 1996.

Chapter 4

1. The 1971 Paris Conference on Standardization in Human Genetics. The March of Dimes.

Chapter 5

1. Sherr, C.J., Roberts, J.M. Inhibitors of mammalian G1 cyclin dependent kinases. Genes and Development, 9:1149. 1995.

Chapter 7

1. Baringa, M., A New Twist to the Cell Cycle. Science, 269:631. 1995.
2. Becker, W.M., Reece, J.B., Poenie, M.F. The World of the Cell, 3rd ed. Menlo Park, California. Benjamin/Cummings, 1996.

Chapter 8

1. Becker, W.M., Reece, J.B., Poenie, M.F. The World of the Cell, 3rd ed. Menlo Park, California. Benjamin/Cummings, 1996.

Chapters 9, 10, 11, 12

1. Mathews, C., Van Holde, K., 2nd ed. Biochemistry, Menlo Park, CA, Benjamin/Cummings Publishing Co. 1996.

Chapters 13, 14, 15, 16

1. Bischoff, J., Kirn, D., Williams, A., Hiese, C. et al. An adenovirus mutant that replicates selectively in p53-deficient cells. Science. 274:373–376. 1996.
2. Kastan, M.B., Zhan, Q., El-Deiry, W. S. et al. A mammalian cell cycle checkpoint pathway utilizing p53 and GADD45 is defective in ataxia-telangiectasia. Cell, 71:587–597. 1992.
3. Ko, L. J. and Prives, C. 1996. p53: puzzle or paradigm. Genes and Develop., 10:1054–1072.
4. Leach, F.S., Nicolaides, N.C., Papadopoulos, N. et al. Mutations of a *Mut S* homolog in hereditary nonpolyposis colorectal cancer. Cell, *75*: 1215–1225. 1993.
5. Levine, A. 1997. p53, the cellular gatekeeper for growth and division. Cell. 88:323–331.
6. Loeb, L.A. Microsatellite instability: Marker of a mutator phenotype in cancer. Cancer Research, *54*: 5059–5063. 1994.
7. Modrich, P. Mismatch repair, genetic stability and cancer. Science 266: 1959–1960. 1994.
8. Sancar, A. Excision repair in mammalian cells. J. Biological Chemistry, *270*: 15915–15918. 1995.
9. Warren, S. The expanding world of trinucleotide repeats. Science, *271*: 1374–1375. 1996.

Chapters 17, 18, 19, 20, 21, 22, 23, 24

1. Knudson, A.G. Mutation and Cancer. A statistical study of Retinoblastoma. PNAS 68:820–823. 1971.
2. Lasko, D., Cavenee, W., and Nordenskjold, M. Loss of Constitutional Heterozygosity in Human Cancer. Ann. Rev. of Genetics, 25:281–314, 1991.
3. Levine, A. The tumor suppressor genes. Ann Rev. Biochemistry, 62:623–651, 1993.
4. Newsham, I., Hadjistilianou T. and Cavenee W. Retinoblastoma, in The Metabolic and Molecular Bases of Inherited Diseases. 7th ed. Eds. Scriver, C., Beaudet, A., Sly, W. and Valle, D. New York, NY: McGraw Hill Inc. 1995, pp 613–642.
5. Nikolov, D.B. and Burley, S. RNA polymerase II transcription initiation: A structural view. Proc Natl. Acad Sci., 94:15–22, 1997.
6. Orphanides, G., Lagrange, T. and Reinberg, D. The general transcription factors of RNA polymerase II. Genes and Development, 10:2657–2683. 1996.
7. Park, M. 1995. Oncogenes: Genetic Abnormalities of Cell Growth, in The Metabolic and Molecular Bases of Inherited Disease, 7th ed. Eds. Scriver, C., Beaudet, A., Sly, W., and Valle, D. New York, NY: McGraw Hill Inc. 1995, pp 589–611.
8. Shao, Z., Seigert, J., Roppert, S. Robbins, P.D. Rb interacts with TAFII250/TFTIID through multiple domains. Oncogene. 15:385–392, 1997.
9. Struhl, K., Chromatin Structure and RNA polymerase II Connection: Implications for transcription. Cell, 84:179–182, 1996.
10. Tjian, R. Molecular Machines that control genes. Scientific American, 272:54–61, 1995.
11. Weatherall, D.J., Clegg J.B., Higgs D.R. and Wood W.G. The Hemoglobinopathies, in The Metabolic and Molecular Bases of Inherited Diseases. 7th ed. Eds. Scriver, C., Beaudet, A. Sly, W and Valle, D. New York, NY: McGraw Hill Inc. 1995, pp 3417–3484.
12. Weinberg, R. Tumor suppressor genes. Science. 254:1138–1146, 1991.

Chapter 25

1. Bennett, R.L., Steinhaus, K.A., Uhrich, S.B., et al: Recommendations for standardized human pedigree nomenclature. Am. J. Hum Genet. 56:745–752, 1995.
2. Marazita, ML. Inheritance patterns in human populations. Medical Molecular Genetics. Fence Creek, Madison, Connecticut, 1998, pp 229–257.

Chapters 26, 27, 28

1. Beutler, E. Gaucher Disease. In Advances in Genetics, Volume 32, J.C. Hall and J.C. Dunlap, eds. Academic Press, NY. 1995. pp 17–49.
2. Eisensmith, R.C. and Woo, S. Molecular Genetics of Phenylketonuria. In Advances in Genetics, Volume 32, Hall J.C. and Dunlap, J.C. eds. Academic Press, NY. 1995. pp 199–271.
3. Grabowski, G.A. Gaucher Disease: Enzymology, Genetics and Treatment. In Advances in Human Genetics, Volume 21, Harris, H. and Hirschhorn, K., eds. Plenum Press, NY. 1992, pp 377–441.
4. Gravel, R.A., Clarke, J., Kaback, M.M., et al. The GM2 Gangliosidoses. in The Metabolic and Molecular Bases of Inherited Diseases. 7th ed. Eds. Scriver, C., Beaudet, A. Sly, W. and Valle, D. New York, NY: McGraw Hill Inc. 1995, pp 2839–2879.
5. Scriver, C.R., Kaufman, S. Eisensmith, R., Woo S. The Hyperphenylalaninemias. in The Metabolic and Molecular Bases of Inherited Diseases. 7th ed. Eds. Scriver, C., Beaudet, A. Sly, W. and Valle, D. New York, NY: McGraw Hill Inc. 1995, pp 1015–1075.
6. Tsui, L-C. and Buchwald M. Biochemical and Molecular Genetics of Cystic Fibrosis. In Advances in Human Genetics, Volume 20, Harris, H. and Hirschhorn, K. eds. Plenum Press, NY., 1991. pp 153–266.
7. Welsh, M.J., L-C Tsui, T.F. Boat and A.L. Beaudet. Cystic Fibrosis. in The Metabolic and Molecular Bases of Inherited Diseases. 7th ed. Eds. Scriver, C., Beaudet, A. Sly, W. and Valle, D. New York, NY: McGraw Hill Inc. 1995, pp 3799–3876.

Chapter 29

1. Marazita, ML. Inheritance patterns in human populations. Medical Molecular Genetics. Fence Creek, Madison, Connecticut, 1998, pp 229–257.

Chapters 30, 31, 32, 33, 34, 35

1. Byers, P.H. Disorders of Collagen Biosynthesis and Structure. in The Metabolic and Molecular Bases of Inherited Diseases. 7th ed. Eds. Scriver, C., Beaudet, A. Sly, W. and Valle, D. New York, NY: McGraw Hill Inc. 1995, pp 4029–4077.
2. Goldstein, J. Hobbs, H. and Brown, M. Familial Hypercholesterolemia. in The Metabolic and Molecular Bases of Inherited Diseases. 7th ed. Eds. Scriver, C., Beaudet, A. Sly, W. and Valle, D. New York, NY: McGraw Hill Inc. 1995, pp 1981–2030.
3. Harper, P. Myotonic Dystrophy and other Autsomal Muscular Dystrophies. in The Metabolic and Molecular Bases of Inherited Diseases. 7th ed. Eds. Scriver, C., Beaudet, A. Sly, W. and Valle, D. New York, NY: McGraw Hill Inc. 1995, pp 4227–4251.
4. Hobbs, H.H, Russell, D.W., Brown, M. and Goldstein, J. The LDL Receptor Locus and Familial Hypercholesterolemia. Ann. Rev. of Genetics, 1990, 24:133–170.
5. The Huntington's Disease Collaborative Research Group. A Novel Gene Containing a trinucleotide repeat that is expanded and unstable on Huntington's Disease Chromosomes. Cell, 72:971–983, 1993.
6. Kahn, P. Coming to Grips with Genes and Risk. Science, 274:496–498. 1996.
7. Knudson, A.G. Mutation and Cancer. A statistical study of Retinoblastoma. PNAS 68:820–823. 1971.
8. Lasko, D., Cavenee, W. and Nordenskjold, M. Loss of Constitutional Heterozygosity in Human Cancer. Ann. Rev. of Genetics, 25:281–314, 1991.
9. Malkin, D. Frederick, P. Li, L. et al. Germ line p53 mutations in a familial syndrome of breast cancer, sarcomas and other neoplasms. Science. 250: 1233–1238, 1990.
10. Nicholls, A. C. Collagen Disorders. In. The Inherited Metabolic Diseases, Holton, J. ed. 1994. Churchill Livingstone, London. pp 379–420.
11. Peifer, M. β-catenin as oncogene: The smoking gun. Science, 275:1752–1753, 1997.
12. Resnik, E. β-catenin—one player, two games. Nature genetics, 16:9–11, 1997.
13. Roa, B., Boyd, A., Volcik et al. Ashkenazi Jewish population frequencies for common mutations in BRCA1 and BRCA2. Nature Genetics, 14:185–187. 1996.
15. Wallace, M., Collins, F.S. Molecular genetics of von Recklinghausen neurofibromatosis. in Advances in Human Genetics, Vol 20. Eds. Harris, H. Hirschhorn, K. Plenum Press, NY. 1991, pp 267–307.
16. Wang, J., Pegoraro, E., Menegazzo, E., Gennarelli, M., Hoop, R., Angeline, C., and Hoffman, E. Myotonic Dystrophy: evidence for a possible dominant-negative RNA mutation. Human Molecular Genetics. 4:599–606, 1995.

Chapter 36

1. Hoffman, E.P. and L.M. Kunkel. Dystrophin Abnormalilties in Duchenne/Becker Muscular Dystrophy. Neuron 2:1019–1029. 1989.
2. Marazita, M.L. Inheritance patterns in human populations. Medical Molecular Genetics. Fence Creek, Madison. Connecticut, 1998, pp 229–257.

Chapter 37

1. Nolin, S., Lewis, F., Ye, LL. et al. Familial transmission of the FMR1 CGG repeat. Am. J. Hum Genet. 59:1252–1261. 1996.

Chapter 38

1. Bindoff, L. and Turnbull, D. M. Defects of the mitochondrial respiratory chain. In The Inherited Metabolic Diseases, Holton, J. ed. 1994. Churchill Livingstone, London. pp 265–295.
2. Johns, D. The other human genome: Mitochondrial DNA and disease. Nature Medicine 2:1065–1068. 1996.
3. Shoffner, J. and Wallace, D. Oxidative Phosphorylation Diseases. in The Metabolic and Molecular Bases of Inherited Diseases. 7th ed. Eds. Scriver, C., Beaudet, A., Sly, W. and Valle, D. New York, NY: McGraw Hill Inc. 1995, pp 1535–1609.

Chapters 39 and 40

1. Marazita, M.L. Inheritance patterns in human populations. Medical Molecular Genetics. Fence Creek, Madison, Connecticut, 1998, pp 229–257.

Chapter 42

1. Miller, R.G. and Hoffman, E.P. Molecular Diagnosis and Modern Management of Duchenne Muscular Dystrophy. Pediatric Neurogenetics 12:699–724, 1994.
2. Schwartz, L.S., Tarleton, J., Popovich, B., Seltzer, W. K., and Hoffman, E.P. Fluorescent Multiplex Linkage Analysis and Carrier Detection for Duchenne/Becker Muscular Dystrophy. Am. J. Hum Genet. 51:721–729, 1992.

Chapters 47, 48, 49, 50, 51

1. Collins, F. Positional cloning moves from Perditional to Traditional. Nature, Genetics, 9:347–350. 1995.
2. Ellis, N. A., Groden, J., Ye, T-Z., et al. The Bloom Syndrome Gene Product is Homologous to Rec Q Helicase. Cell. 83:655–666, 1995.
3. Gray, J.W. and Langlois, R. G. Chromosome Classification and Purification Using Flow Cytometry and Sorting. Ann. Rev. Biophys. Biophys. Chem. 15:195–235. 1986.
4. Hoffman, E.P. and L.M. Kunkel. Dystrophin Abnormalities in

Duchenne/Becker Muscular Dystrophy. Neuron 2:1019–1029, 1989.
5. Hudson, T. J., Stein, L.D., Gerety, S.S. et al. An STS-based Map of the Human Genome. Science. 270:1945–1954. 1995.
6. Koenig, M., Hoffman, E.P., Bertelson, C.J. et al. Complete cloning of DMD cDNA and Preliminary Genomic Organization of the DMD Gene in Normal and Affected Individuals. Cell 50:509–517. 1987.
7. Leach, F. S., Nicolaides, N.C., Papadopoulos, N. et al. Mutations of a *mut S* Homologue in HNPCC. Cell 75:1215–1225. 1993.
8. Marazita, M.L. Inheritance patterns in human populations. Medical Molecular Genetics. Fence Creek, Madison, Connecticut, 1998, pp 229–257.
9. Miller, R.G. and Hoffman, E.P. Molecular Diagnosis and Modern Management of Duchenne Muscular Dystrophy. Pediatric Neurogenetics 12:699–724, 1994.
10. Murray, J. C., Buetow, K.H., Weber, J. L., et al. A Comprehensive Human Linkage Map with Centimorgan Density. Science. 265:2049–2055. 1994.
11. Peltomaki, P., Aaltonen, L.A., Sistonen, P. et al. Genetic Mapping of a Locus Predisposing to Human Colorectal Cancer. Science 260:810–812. 1993.
12. Shows, T.B. and Sakaguchi, A. Y. Gene Transfer and Gene Mapping in Mammalian Cells in Culture. In Vitro. 16:55–76. 1980.

Chapters 52, 53, 54

1. Hook, E.B.: Chromosome abnormalities: prevalence, risks and recurrence. In Prenatal Diagnosis and Screening. Edited by Brock D.J.H., Rodeck, C.H., Ferguson-Smith, M.A., Edinburgh, UK: Churchill Livingstone, 1992, pp 351–392.
2. Mowery-Rushton, P.A., Surti, U. Phenotypic expression of chromosome abnormalities. In Medical Molecular Genetics. Fence Creek, Madison, Connecticut, 1998, pp 193–227.

Chapter 56

1. Buntinx, I.M., Hennekam, R.C.M., Brouwer, O.F. et al: Clinical profile of Angelman syndrome at different ages. Am. J. Med Genet 56:176–183, 1995.
2. Dittrich, B. Buiting, D. Korn, B. et al: Imprint switching on human chromosome 15 may involve alternate transcripts of the SNRPN gene. Nat Genet 14:163–170, 1996.
3. Kishino, T., Lalande, M., Wagstaff, J: UBE3/E6-AP mutations cause Angelman syndrome. Nat Genet 15:770–773, 1997.
4. Ledbetter, D.H., Ballabio, A.: Molecular cytogenetics of contiguous gene syndromes: mechanisms and consequences of gene dosage imbalance. In The Metabolic and Molecular Bases of Inherited Diseases. 7th ed. Eds. Scriver, C., Beaudet, A., Sly, W. and Valle, D. New York, NY: McGraw Hill Inc. 1995, pp 811–839.
5. Mowery-Rushton, P.A., Surti, U. Phenotypic expression of chromosome abnormalities. In Medical Molecular Genetics. Fence Creek, Madison, Connecticut, 1998, pp. 193–227.
6. Williams, C.A., Angelman, H. Clayton-Smith, J. et al: Angelman syndrome: consensus diagnostic criteria. Am. J. Med. Genet. 56:237–238, 1995.

Chapter 57

1. Kuroda, M.I., Meller, V.H.: Transient Xist-ence. Cell 91:9–11, 1997.
2. Mowery-Rushton, P.A., Surti, U. Phenotypic expression of chromosome abnormalities. In Medical Molecular Genetics. Fence Creek, Madison, Connecticut, 1988, pp. 193–227.
3. Sheardown, S.A., Duthie, S.M., Johnston, C.M. et al. Stabilization of Xist RNA mediates initiation of X chromosome inactivation. Cell 91:99–107, 1997.

Chapter 58

1. Mowery-Rushton, P.A., Surti, U. Phenotypic expression of chromosome abnormalities. In Medical Molecular Genetics. Fence Creek, Madison, Connecticut, 1998, pp 193–227.
2. Netley, C.T.: Summary overview of behavioral development in individuals with neonatally identified X and Y aneuploidy. In Prospective Studies on Children with Sex Chromosome Aneuploidy. Edited by Ratcliffe S.G., Paul, N., New York, NY; Alan R. Liss 1986, pp 293–306.
3. Saenger, P. Turner syndrome. N Engl J Med. 335:1749–1754, 1996.

Chapter 59

1. Mowery-Rushton, P.A., Surti, U. Phenotypic expression of chromosome abnormalities. In Medical Molecular Genetics. Fence Creek, Madison, Connecticut, 1998, pp 193–227.
2. Mitelman, F.: Catalog of Chromosome Aberrations in Cancer, 5th ed. New York, NY: Wiley-Liss, 1995.
3. Rowley, J.D.: The Philadelphia chromosome translocation, A paradigm for understanding leukemia. Cancer 65:276–293, 1973.
4. Sandberag, A.A.: The Chromosomes in Human Cancer and Leukemia, 2nd ed. New York, NY: Elsevier, 1990.

Chapter 60

1. Ferguson, J.E., Vick, D.J., Hogge, J.S. et al: Transcervical chorionic villus sampling and amniocentesis: a comparison of reliability, culture findings and fetal outcome. Am J Obstet Gynecol 163:926–931, 1990.
2. Hogge, W.A. Clinical Aspects of Medical Genetics. In Medical Molecular Genetics. Fence Creek, Madison, Connecticut, 1998, pp 315–331.
3. Rhoads, G.R., Jackson, L.G., Schlesselman, S.E. et al: The safety and efficacy of chorionic villus sampling for early prenatal diagnosis of cytogenetic abnormalities. N. Engl J. Med 320:609–617, 1989.

Chapter 61

1. Crystal, R. G. Transfer of genes to humans: early lessons and obstacles to success. Science, 270:404–410. 1995.
2. Gao, X. Huang, L. Cationic liposome-mediated gene transfer. Gene Therapy, 2:710–722. 1995.
3. Morgan, R. A. Anderson, W. F. Human gene therapy. Annual Reviews of Biochemistry. 62:191–217. 1993.
4. Verma, I.M., Somia, N. Gene therapy—promises, problems and prospects. Nature. 389:239–242. 1997.

Chapter 62

1. Crystal, R. G. Transfer of genes to humans: early lessons and obstacles to success. Science, 270:404–410. 1995.
2. Salmons, B. and Gunzburg, W. Targeting of retroviral vectors for gene therapy. Human Gene Therapy. 4:129–141. 1993.

Chapter 63

1. Curiel, D. Adenovirus facilitation of molecular conjugate-mediated gene transfer. Progress in Medical Virology 40:1–18. 1993.
2. Harrington, J., Van Bokkelen, G. Mays, R. Gustashaw, K. and Willard, H. Formation of de novo centromeres and construction of first-generation human artificial chromosomes. Nat Genet 15:345–355, 1997.

Chapter 64

1. Blaese, R., Culver, K., Miller, A., Carter, C. et al. T lymphocyte-directed gene therapy for ADA-SCID. Initial trial results after 4 years. Science 270:475–480. 1995.

2. Grossman, M. et al. A pilot study of ex vivo gene therapy for homozygous familial hypercholesterolemia. Nature, Med. 1:1148–1154. 1995.
3. Haecker, S., Stedman, H., Balice-Gordon, R., Smith, D. et al. In vivo expression of full-length human dystrophin from adenoviral vectors deleted of all viral genes. Human Gene Therapy, 7:1907–1914. 1996.
4. Knowles, M. et al. A controlled study of adenoviral-mediated gene transfer in the nasal epithelium of patients with cystic fibrosis. New England J. of Med. 333:823–831. 1995.
5. Pelletier, J.P., Caron, J., Evans, C.H., Robbins, P.D. et al. In vivo suppression of early experimental osteoarthritis by IL-1Ra using gene therapy. Arthritis Rheum 40:1012–1019.

Chapter 65

1. Barlow, C. Hirotsune, S., Paylor, R., Liyanaage, M. et al. Atm-deficient mice: a paradigm of Ataxia telangiectasia. Cell 86:159–171, 1996.
2. Bedell, M., Largaespada, D., Jenkins, N. and Copeland, N. Mouse models of human disease: Part II. Recent progress and future directions. Genes and Development. 11:11–43. 1997.
3. Capecchi, M. Targeted gene replacement. Scientific American 271:52–59, 1994.
4. Colledge, W., Abella, B., Southern, D., Ratcliff, R. et al. Generation and characterization of a delta F508 cystic fibrosis mouse model. Nature Genetics. 10:445–448. 1995.
5. Fero, M., Rivkin, M., Tasch, M., Porter, P. et al. A syndrome of multiorgan hyperplasia with features of gigantism, tumorigenesis and female sterility in p27 (kip1) deficient mice. Cell 85:733–744. 1996.
6. Hakem, R., de la Pompa, J., Sirard, C., Mo, R. et al. The tumor suppressor gene BRCA1 is required for embryonic cellular proliferation in the mouse. Cell. 85:1009–1023. 1996.
7. Wilmut, I., Schnieke, A., McWhir, J., Kind, A., and Campbell, K. Viable offspring derived from fetal and adult mammalian cells. Nature: 385:810–813. 1997.

Index